SOLDIER OF FORTUNE
GUIDE TO HOW TO BECOME A
MERCENARY

SOLDIER OF FORTUNE
GUIDE TO HOW TO BECOME A
MERCENARY

BARRY DAVIES

Skyhorse Publishing

Disclaimer: This book is intended to offer general guidance relating to the mercenary occupation, one of the most dangerous occupations that exists. It is sold with the understanding that every effort was made to provide the most current and accurate information. However, errors and omissions are still possible. Any use or misuse of the information contained herein are solely the responsibility of the user, and the author, publisher, and licenser make no warrantees or claims as to the truth or validity of the information. The author, publisher, and licenser shall have neither liability nor responsibility to any person or entity with respect to any loss or damage caused, or alleged to have been caused, directly or indirectly, by the information contained in this book.

Warning: This book includes graphic images taken in war zones throughout the world. Viewer discretion is advised.

Skyhorse Publishing books may be purchased in bulk at special discounts for sales promotion, corporate gifts, fund-raising, or educational purposes. Special editions can also be created to specifications. For details, contact the Special Sales Department, Skyhorse Publishing, 307 West 36th Street, 11th Floor, New York, NY 10018 or info@skyhorsepublishing.com.

Skyhorse® and Skyhorse Publishing® are registered trademarks of Skyhorse Publishing, Inc.®, a Delaware corporation.

Visit our website at www.skyhorsepublishing.com.

10 9 8 7 6 5 4 3 2 1

Library of Congress Cataloging-in-Publication Data

Davies, Barry, 1944-
Soldier of fortune guide to how to become a mercenary / Barry Davies. p. cm.
ISBN 978-1-62087-097-6
(pbk. : alk. paper)
1. Private military companies–Vocational guidance. 2. Private security services–Vocational guidance. 3. Mercenary troops–Vocational guidance. 4. Mercenary troops–Handbooks, manuals, etc. I. Title. II. Title: Soldier of fortune.
UB149.D38 2013
355.3'54023--dc23
                    2012035543

Printed in China

In memory of William "Bill" Cain, who died in 2012 in Hereford

CONTENTS »

INTRODUCTION »

I have read many accounts of mercenaries and soldiers of fortune, and while some are quite interesting, they tend to lack any true inside knowledge of what it takes to be a modern-day private contract officer. Therefore, I would like to start this book off in the right context. The word *mercenary* appears on the cover of this book, but to be honest, the day of the mercenaries—or the soldier of fortune—has long since passed. Personally, I would like to call such men and women "private contract officers," as this is more in keeping with their role in the modern world. It is true that there are still mercenaries out there, but these are mainly idiots with guns who have no idea of what is happening in the real world. The old mercenary bands have all but gone, replaced by what is now called private military contractor (PMC) or a private security company (PSC).

However, I must hastily draw a line and explain the vast difference between private military contractors and mercenaries. PMCs are subject to government regulations and prosecution to ensure their standards of conduct are acceptable and comply with international law. These standards are not readily seen by some, and the press often describes many PMCs as mercenaries merely because they are armed security services being contracted, rather than regular troops. The media, often in the heat of war, which is always foggy at best, find the distinction between PMCs and regular military officers sometimes distorted.

Finally, I will be the first to admit that there is no way this book can teach you to be a mercenary, soldier of fortune, or a private contract officer. The best I can offer is to let you know what you are getting into and the skills you will need to acquire. If you have been a professional soldier, you will understand and appreciate what I have written; if you are a civilian starting from scratch, read this book cover to cover and then go and find yourself a nice safe civilian job.

In 2013, the private military industry has become a reality. The subcontracting of military-related logistics, security, training, and support is due entirely to the decrease in military commitment and the general downsizing of regular armies. For the past twenty-five years, war in one form or another has been continuous—with both the United States and Great Britain actively participating. It is not just the larger wars, such as those in Iraq, Afghanistan, and Libya that drain manpower, but the standing commitments demanded on both nations. We have a tendency to forget about South Korea, where America has around 25,000 troops permanently based, or Northern Ireland and the Falklands, where some 300 British troops are stationed.

Then there are the smaller conflicts that no one is bothered about, like the genocide in Rwanda. It is true that the United Nations (UN) did send troops under the United Nations Assistance Mission for Rwanda (UNAMIR), but in typical UN style, the commanders on the ground were hampered by the UN committee, which as always, did not want to upset some of its member nations. I will take this opportunity to call the UN gutless, for while they have excellent forces at their disposal, they are refrained by the UN from taking serious action. This has happened in so many small conflicts that it begs the question: "Why do we bother with the UN?" Look what happened in Rwanda while the UN stood idly by. Secretary-General Kofi Annan expressed his "deep remorse" over the UN's failure to halt the massacre of 800,000 men, women, and children.

Rwanda has two main ethic groups: the Hutu (around 80% majority) and the Tutsi. In 1993, businessmen close to General Habyarimana imported over half a million machetes. These were cheaper than guns and intended for Hutus' use in killing Tutsis. In general, the Tutsis are lighter colored than the Hutus and thus easier to distinguish. In addition, both Hutu and Tutsi were forced to carry ID cards.

The killing started and was fueled by a media campaign, which encouraged the Hutu gangs to seek out and kill Tutsis. Most of the victims were killed in their own villages or in towns—often by their neighbors and fellow villagers. It was not unusual for a Hutu gang to enter a village and hack every man, woman, and child to death.

On April 9, 1994, UN observers entered a Polish church in the village of Gikondo and found around 110 Tutsis, most of which were children. They had been murdered by Interahamwe militia (government-organized Hutu gangs), who were acting under the protection of the Hutu presidential bodyguard. On the same day as this barbaric incident took place, over 1,000 heavily armed and well-trained European troops arrived to escort European civilians out of the country. In my book, that number of European troops could have taken on and stopped any more killings—but they did not.

Author's Note: It is my experience that it's best to avoid doing any contract work in Africa. I have several friends working there, and while they are making good money, it is dangerous—especially when you venture outside the safe areas. The African rebels' mentality is harsh: They have little fear of dying and no compunction in killing someone. In addition, they have an extraordinary supply of firepower, which usually ends up in the hands of illiterate, drug-induced, young morons. They enjoy—or even find it entertaining—to torture, mutilate, and rape their captives (male and female). While there may be rich contracts being offered, it is not worth it . . . so stay out of Africa.

However, the multiple large-scale military operations in the Middle East have forced governments to find much needed additional resources to conduct these operations, specifically in Iraq and Afghanistan. Today it has become politically acceptable—and financially feasible—to replace regular military with PMCs. It is also desirable as, apart from reducing the demands on the military, PMCs offer a range of skills outside the military sphere.

One of the original mercenary groups was Executive Outcomes (EO) in South Africa. Executive Outcomes was one of the first PMCs to emerge, and its success set the example for other PMCs today. EO was one of the first companies in modern times that hired themselves out to a sovereign state in order to do their military work. They were a small, private foreign army, armed and actually fighting for the Angolan government. They did a great job, worked to the contract they signed up for, acted with self-discipline, and got the job done with minimum loss of life.

By far the largest of the PMCs today is the U.S. company, Blackwater. Their rise to power and their decline after an incident in Iraq highlights the good, the bad, and the downright ugly of the private security industry. Needless to say, it is difficult to control some of the idiots that get contract work with large PMCs, and I am glad to see that Blackwater has now come back even stronger due to its excellent leadership.

SOME FRIENDLY ADVICE

Working for a Private Military Company (PMC) or Private Security Company (PSC) has to be one of the most dangerous jobs in the world. You will first think about the money you could be earning fighting as a soldier of fortune in a war zone, but my advice is to think about your family and your chances of survival. In Iraq alone, over 1,000 contractors died and thousands were wounded between 2003 and 2011. . . . And as of the March 16, 2012, more than 200 private contractors died in Afghanistan, of which some forty-two were working for PMCs, and most were American (see Annex A).

At the end of the first Gulf War, there was a massive reduction in regular military forces, and many of those who had to leave were still looking to do what they did best. Some PMCs had already been working very successfully since the 1970s, but the occupation of Iraq saw an explosion of contract work being offered. Currently, the United States Defense Department employs over 196,000 contractors in Iraq and Afghanistan in one role or another; this number even outnumbers the regular troops deployed to the same locations.

While many do construction or security work, only about 10% are used in a tactical armed role, but they all have one thing in common: They are all working in a war zone.

The modern soldier of fortune is now called a security or contract officer, contractor, or consultant, and he or she will work for a bona fide company that is employed by governments or other legitimate agencies as well as individuals to carry out a wide variety of tasks. PMCs employ a large selection of highly trained individuals with specialist skills. Most of their employees will be retired military or police, or from a reputable security agency. However, PMCs also require non-military people with specialist skills, such as construction engineers, doctors, and non-military pilots.

I think it is fair to say that the PMC (from this point, take it that PMCs and PSCs are one and the same to save the repetition) is the modern successor to the days of the old-fashioned mercenary, in the sense that the old model of "guns for hire" has developed into a legalized and well-governed profession. They certainly still work in the same theatre, but the type of work has shifted more towards supplying legitimate services, such as providing military training, close protection (CP), and logistics among other things. PMCs are also often authorized to accompany regular armed forces into the battle zone and provide such paid services that may be deemed necessary. These can be anything from the close protection of media crews or senior members of nongovernmental organizations (NGOs) to the movement of vital goods and equipment through hostile territories.

During the second Iraq war, one British military contractor supplied an armed escort for the Japanese military, escorting their convoys as they moved back and forth between Kuwait and Iraq: They were responsible for protecting the army that was there to protect the Iraqis. Despite the rundown in operations in Iraq, the Private Security Industry (PSI) is booming, and while many companies are American, British companies also make up a large proportion.

In writing this book, I do not intend to waste pages explaining the law and control of mercenaries or the impact some PMCs have made. Instead, this book is about what you need to know should you want to become a soldier of fortune. With that said, there is a clear definition of what a mercenary is and what constitutes a contract officer working for a PMC.

Mercenary: The Geneva Convention (1977) from Article 47 of Protocol I.

1. A mercenary shall not have the right to be a combatant or a prisoner of war.

2. A mercenary is any person who:

(a) is especially recruited locally or abroad in order to fight in an armed conflict;

(b) does, in fact, take a direct part in the hostilities;

(c) is motivated to take part in the hostilities essentially by the desire for private gain and, in fact, is promised, by or on behalf of a Party to the conflict, material compensation substantially in excess of that promised or paid to combatants of similar ranks and functions in the armed forces of that Party;

(d) is neither a national of a Party to the conflict nor a resident of territory controlled by a Party to the conflict;

(e) is not a member of the armed forces of a Party to the conflict; and

(f) has not been sent by a State which is not a Party to the conflict on official duty as a member of its armed forces.

In the end, it all comes down to the fact that true mercenaries have no law; they are not restricted to, or forced to abide by, international laws or government regulations. If they have any standards, it is to the paymasters that employ them, added to which under the United

Nations, you do not come under the protection of the Geneva Convention.

Author's Note: A classic mercenary is best described by those soldiers who fought in Africa during the 1960s. Mike Hoare—real name: Thomas Michael Hoare—made his name as a mercenary in the Congo during that time. His exploits were well publicized in the press, and they even made a film (*Wild Geese*) based on his campaigns. Born in Dublin, in the Republic of Ireland, he served with the British army during World War II as a captain but afterwards immigrated to South Africa where he made a living offering safaris to tourists.

In 1961 he raised a mercenary unit known as "4 Commando," which employed mainly Belgian ex-paratroopers to work in the breakaway province of Katanga. Later, in 1964, Hoare returned to the Congo with another unit, "5 Commando," with the primary aim of extracting European civilian workers and missionaries from Stanleyville, where their lives were under threat from the Simba rebellion. The Simbas were a violent, drug-crazed force, hostile to the Congolese Central Government, who subjected most of their captives to being hacked to death with a machete. The fighting at the time was vicious, and many of the Simbas believed themselves to be invincible to bullets, forcing the mercenaries to shoot them in the head in order to stop them. Hoare was also a strict disciplinarian amongst his own men and once had the large toe cut off one of his soldiers for raping and killing a young girl.

His mercenary career hit the headlines again in 1981, when he led a team into the Seychelles. The idea was to lead a force on behalf of ex-president James Mancham in order to stage a coup. This time his team consisted of South African Special Forces, some Rhodesian soldiers, as well as members from his ex-Congo days. Disguised as tourists, they arrived at Mahe Airport, but a customs

Bob Denard, the French soldier-of-fortune whose near mythical involvement in African wars since the 1960s made him one of the world's most famous mercenaries, died at the age of 78. He was with Hoare when they went to rescue white civilians stranded in Stanleyville.

officer spotted an AK-47 in one of the bags and all hell broke loose. After a brief gun-battle at the airport, most of the mercenaries escaped by hijacking an Air India jet, which happened to be on the runway. Only one mercenary was killed during the skirmish, but seven others (six men and one woman) were captured. The Seychelles government tried the men between June and July of 1982 (the charges against the woman were dropped), and four of the six were sentenced to death. However, after serious negotiations, all were eventually returned to South Africa in mid-1983, where Hoare and his mercenaries were immediately arrested and tried. This was not, however, for having attempted to organize a coup in a foreign country but for specific offences under the Civil Aviation Offences Act of 1972. Hoare received ten years, and many of his men served between two and five. It is true to say that Mike Hoare glamorized the mercenary trade for a whole generation, but the incident in the Seychelles illustrates just how easy it is to make a mistake and pay dearly for it.

International Code of Conduct (ICoC) for Private Security Service Providers

Although PMCs, operating from a variety of countries, have been around for several years, there used to be no real control or rules governing their behavior. This was the case until the International Code of Conduct (ICoC) for Private Security Service Providers was established and some form of regulatory body at last came into being. The process started with a draft proposal being put forward on May 23, 2012, in London followed by a discussion of the charter in Washington on May 28. By April 1, some 357 companies had signed up from over fifty-five different countries.[1]

The ICoC is a set of principles for private military and security providers, formed through a multi-stakeholder initiative organized by the Swiss government. It is simply a code that emphasizes and articulates the obligations of private security companies and providers. It contains a set of rules, which they have promised they will all abide by, in particular with regard to international humanitarian law and human rights law. The ICoC also sets the foundation for developing an institutional framework to provide meaningful and independent oversight of, and accountability to, the ICoC.

* * *

So you wake up one morning and decide you want to join the elite and work as a contract officer or soldier of fortune and earn a lot of money. Unfortunately it's not quite that simple, as you will need a range of specialist skills. If you are retired from the military or police force, you will already have some of these skills; if not, I would suggest you go back to bed. To be perfectly honest, no matter what I write in this book, it will not make you an expert in military or security craft: If you think that, you will just be deluding yourself. The best I can do is to explain what is required.

---

[1] www.icoc-psp.org/

First and foremost, you need to know what is happening in the world. You need to know that while the majority of people travel freely around our planet, there are areas where it is simply not safe to go. These areas are normally war or conflict zones, places where the gun and bomb rule and deal out justice. While the regular military operates in these war zones, there is a certain amount of order and protection.[2] In other conflicts, where there are no regular or peace-keeping soldiers—just rebel factions fighting each other (the trouble in Syria while this book is written is a good example)—the conditions are unsafe to the extreme. These are places you definitely do not want to go to. If you do venture into the latter, then the PMC you are working for should be paying you an awful lot of money, something around $1,000+ a day. Even so, a professional retired soldier would still have to weigh up the risk.

While the growth of the PSI has increased at an astonishing rate, PMCs and PSCs take only the best manpower available, as their reputation depends on it. If you served in a well-known military regiment or corps, then you will easily find work as a contract officer. How high up the ladder you go will depend very much on the unit you served with. If you were a member of Delta Force, SEAL Team 6, or the British SAS or SBS, you should be able to literally walk into any company, as most of the better PMCs are run by retired members. The reason for this is simple: They have the skills, most are battle hardened, and are always very professional. Understand that this is not a slight on the rest of the military, as I have known many excellent soldiers who served in the Royal Marines and other regular regiments. However, Special Forces, by their very nature, learn skill sets beyond the norm, and employees from this background are generally going to receive a higher salary than others. For example, during the second Iraq war, some former British SAS soldiers working in PSCs were on $1,000+ a day, while

---

[2] Afghanistan is a good example of this.

those who had served with a logistics regiment and were employed for escort duties were receiving around $500 a day.

Almost all countries have some form of Special Forces, with some being more prominent than others. Delta Force and SEAL Team 6 are just two of the American units. Having worked with them both, I can honestly say that they are outstanding and deserving of the praise heaped upon them. Likewise, I have worked with the German GSG9 and the French GIGN—again, extremely professional forces. However, I think it would be safe to say that one Special Forces unit stands out from the rest: The British Special Air Service (SAS) is a household name; honored by its peers at home and abroad. In many American action films, the word SAS comes up more and more when talking about elite Special Forces. SAS actions are normally swift and very hard hitting, with their soldiers fading back into obscurity afterwards. What the public have seen of them—like the spectacular hostage-rescue at the London Iranian Embassy—confirms the truth for the many other unseen actions. Yet few see the SAS for what it truly is: 200 men, the best Britain can find, rigorously selected, highly trained, and with a spirit to dare. They will go—willingly—deep behind enemy lines, take on incredible odds, and risk their lives to rescue others or defeat an enemy.

There are many Special Forces units around the world, and although most have an excellent reputation, there are just too many to name. For example, I recently visited with members of the 707 in South Korea: good guys, well trained, and alert to the constant and very real threat from their northern neighbor. However, to give you some idea of what the better known Special Forces do, I have included a brief outline below:

British 22 Special Air Service (SAS)

To join the SAS, one must have spent at least three years in the regular army and then pass selection. There is no other way to say it: The SAS selection course is hard. The basis of the selection system is to ensure that the valuable training time is only spent on those who have

Barry Davies

SAS Selection is tough, with a heavy pack over mountainous terrain and many miles still to go.

a chance of making the grade. It's what makes the SAS so unique: a whole bunch of individuals with the capacity to act as one.

Selection takes place mostly in the mountains of either South Wales or Scotland. The mountain ranges used are not particularly high or technically challenging, but they are treacherous. Exposed and battered by constant weather changes, soldiers are constantly risking death by hypothermia, with many succumbing to this slow death. It is therefore essential that a diligent, self-imposed training schedule be undertaken by the candidate prior to arriving for selection. Stamina and fitness, plus the ability to read an Ordinance Survey map and use GPS, will go a long way to getting the candidate successfully through to Test Week.

Test Week comes at the end of the first phase of SAS selection and is designed to test individual fitness and stamina. It is by far the most physically grueling section of the entire course. At the end of Test Week, the remaining candidates come face to face with the "endurance march." Carrying a rifle and pack weighing 30 kg (66 lb), they are expected to walk 40 km (25 mi) in a time of twenty to twenty-four hours (dependant on the time of year). The route: running up and down the Brecon Beacons.

Continuation training follows, and all the necessary basic skills required of the SAS soldier, allowing him to become a new Squadron member, are taught and practiced. It is back to basics: weapon training, patrolling skills, SOPs (Standard Operational Procedures), escape and evasion exercises, parachuting, and finally five weeks of jungle training. Then for those who cannot swim or drive, there is a crash course in both before the candidate is allowed to enter a Squadron. Finally, when all is done, he will receive his beige beret with its famous winged dagger: As any SAS man will tell you, it is a special moment and a fabulous feeling! From then on, and providing you survive, your world and the rest of your life is one of pure adventure. Plus, when you leave the SAS, there is ALWAYS a job open for you with any PMC.

Delta Force

Delta Force was founded in 1977 by Colonel Charles Beckwith. I knew Charlie well, and he was one mean soldier. He had been attached to the SAS on exchange in 1962 and had served with my regiment in Malaya. When he finally returned to the United States, the first thing he did was to write a report on raising a special unit based along the lines of the SAS. It then took Charlie until 1977 to get his plans for Delta passed by the Chiefs of Staff. Their first hostage rescue operation, however, finished in a total disaster. It started with a coup in Iran, when the Shah was toppled from power to be replaced by the radical Ayatollah Khomeini. The army then became a rabble, known as the Iranian Revolutionary Guards. These militants took over the American Embassy in Tehran and held over a hundred Americans hostage. Delta Force was sent to rescue them. The rescue mission ran into one misfortune after another and ended in failure. This was not the fault of Delta Force but rather just a series of mishaps that sometimes plague military operations. Rumor has it that Charlie broke down in tears at the folly of the administration and lack of support from Central Command.

GIGN

I first met the French GIGN (Groupement d'Intervention de la Gendarmerie Nationale) in the early days when they came to Hereford where I was on the Counter-Terrorist Team. The man then in charge of the GIGN was a tall, good-looking officer named Captain Christian Prouteau. Although the French GIGN had participated in several successful actions, they had to wait until 1994 to get their main chance when an Air France Airbus was hijacked at Algiers Airport.

The first television news of the event reported that the aircraft had been boarded by four Islamic fundamentalists, who had taken the 220 passengers and twelve crewmembers hostage. The aircraft finally landed at Marseilles, where it was immediately surrounded by snipers of the GIGN. By 5:17 p.m., assault units from the GIGN could be seen racing towards the rear of the aircraft, and sniper fire could be heard. The team commander, Denis Favier, together with his second-in-command, Oliver Kim, stormed the front right door. They used normal airport landing steps to gain entry, and despite trouble with the door, the entry, and all the other conditions, the operation was very slick. At the same time, Capt. Tardy led another unit via the right rear door, using the same entry method. Once on board, both groups made their way towards the cockpit, where most of the terrorists had gathered. By 5:39 p.m., the four terrorists lay dead. Nine GIGN were wounded: two quite severely, one lost two fingers of his hand, and one was shot in the foot.

Later reports indicated that the terrorist plan was to fill up the aircraft with fuel and blow it up over Paris; this theory was enforced when twenty sticks of dynamite were found under the front and mid-section seats. I think many Parisians can count themselves lucky that their country has such a force as the GIGN.

Author's Note: Do not worry if you have not been a member of Special Forces. As has been previously stated, there are lots of very good professional military units dotted around the world. And by military, I also include Air Force and Navy units. One true account of someone who succeeded despite being from a non-Special Forces background is outlined below.

Corporal Susan Jones (pseudonym), 42, served in the Royal Air Force (RAF) for ten years, specializing as a survival equipment fitter. She left the forces in April 2008 to start her own catering business: "I owned and ran a fine-dining English restaurant and so was self-employed." However, during her resettlement period, she undertook a close protection (CP) course with a British company: "an excellent course that provided me with the tools to do the job"—after which she "actively pursued employment, personally contacting companies and arranging meetings, compiled an excellent, easy-to-read CV, and traveled far and wide to attend interviews.

"I am now a close protection officer (CPO) in Afghanistan, and a lot of the firearms skills and extraction training I learned while in the military have been very useful to me in this hostile environment. My current position involves the protection and transportation of high-profile clients in and around the city of Kabul. I work in a team and, when moving a client, I may be driver or bodyguard. I am heavily armed with automatic weapons, and we are highly trained in emergency extraction drills, should these be required, complete risk assessments and recon reports for all new venues; we also work with a wide range of comms equipment.

"I love my job. Every day is different and the work is always challenging. The only downside is being away from loved ones. There are similarities between my Service role and my current job in terms of the work and environment, but CP work is far more challenging and rewarding, in my opinion. During my Service career I followed orders and had an awful lot of the decisions made for me. As a CPO (close protection operative) you have to be totally self-reliant, make your own decisions, and ensure that all aspects of the task requirements are covered and fully completed." Asked whether there is there a significant difference in salary, she replied, "Yes, at least double my military salary."

Not a Member of Special Forces or the Military?

I am now going to assume that you have never been in the military, so becoming a member of the Special Forces is way beyond your reach. Well don't be alarmed, as there are many ways civilians can gain valuable military experience and do their country some good in the process. If you join the reserves, you will get all the training you need, at least in all the basics. Even in the reserves you can select a trade and become highly qualified. If you do not want to join the reserves in your own country, you could always go overseas and join a unit such as the French Foreign Legion.

Most countries have reserve armies and organizations that are extremely professional. In the United States, it is possible to join the Army Reserve, while in the United Kingdom, there is the Territorial Army. In both cases and as with many other countries, this simply means that, if accepted, you will be given basic military training and will be required to attend weekend training about one day per month. Additionally, there is generally a requirement to participate in a two-week exercise once a year.

Reserve armies support the regular army in a host of different ways, and with the size of standing armies being reduced, the role of the reserve army is becoming more important. Reserve armies also tend to use the civilian skills of the individual. For example, a nurse in civilian life may well continue to serve in the reserves in a

Barry Davies

Learning basic military skills will include simple things like patrolling in formation and how to handle a weapon.

nursing role. However, individuals with no particular skills may well find themselves learning other military skills, such as how to operate a specialist weapon.

Basic training within any reserve army concentrates on instructing the civilian in the disciplines and regulations of the military, molding him or her into a soldier. This will include, but is not limited to, dress, drill, personal equipment, and weapon handling. It also involves learning about military structure and what part fits in with what role.

What you learn while serving in a reserve army will very much depend on you, the individual. For example, if you want to push yourself, you can always volunteer for additional courses or specialist subjects and training. While most reserve armies restrict the amount of time and days an individual soldier can spend serving, some specialist units have exemptions. Additionally, many reservist units are called up to work in war zones from time to time, and this is where the real experience is gained. If you're really lucky, you may even experience combat; however, you must also remember that if you are unlucky, you may also die in combat.

To enlist in the U.S. Army Reserve, you must be between 18 and 35 years old and must be a U.S. citizen or a legal permanent resident who is physically living in the United States and has a government-issued green card. Additionally, you should have had some formal education, such as a high school diploma (a GED is acceptable), and you must pass a basic fitness test. You will then be sent off to complete a ten-week Basic Combat Course, commonly known as "Boot Camp." The British Territorial Army is very similar in requirement and structure. Joining the reserves not only helps build character but gives you the opportunity to serve your country, make lots a new friends, learn new skills, and have a great time during field exercises.

Barry Davies

Improving your fitness is essential if you wish to work in the security industry.

Improving Your Fitness

Before you join any military organization or PMC, you will need to get yourself in shape. There is only one person who knows the true level of your fitness . . . and that is you. To start, it is vital to be honest with yourself: How fit are you really? Define this and then work out a target. Next, examine your daily routine, diet, and lifestyle, both in work and at home. Drinking and late nights may have to be sidelined for a while. Also, take your age into consideration; you may have great mental determination, but is your body up to it?

Checklist for Fitness

- Do you exercise?
- Could you run ten miles at a steady pace, keeping your breathing under control?
- Could you walk ten miles over hilly country, with a 20 kg (44 lb) backpack?
- Do you drink more than two pints of beer or the alcoholic equivalent each day?
- Do you have high blood pressure?
- Does your diet contain a regular supply of high-fat foods?
- Is your weight comparable with your height?
- Do you smoke?
- Do you take drugs?
- What is your work routine?

Establish a Routine

To reach a good standard of fitness, it is important that you establish a workout routine. Do these workouts at least four times a week for about an hour, and make sure you include a lot of cardio. Vary the length of the routes to take into account your age and physical condition, and once you have reached a standard of fitness, push yourself a little bit further. Walk up hills and jog down them—don't run; it is always a good idea to keep a little bit of energy in reserve. Once a week—perhaps on a Sunday morning—try doubling you distance.

- When bad weather becomes a problem and stops you from training outside, ensure that there is always some form of exercise that you can do indoors. One excellent way of both exercising and controlling your breathing is to work out on a punch bag for half an hour. Alternatively, do simple push-ups followed by shadow boxing.

- If you don't want to spend any money, try walking up a flight of stairs. This is an excellent form of aerobic exercise, but you should start off slow and build up carefully. Start by using a flight of stairs with no more than twenty steps. Walk up at a slow pace and come down at twice the speed. Do this four times, increasing to ten the fitter you get.

Walking

The average person should be able to walk at a pace of 5 mph over flat ground. Start off by walking at your own pace and increase your speed and stride as your body gets warmed up. You do not have to make a special effort to maintain a walking routine; it can become part of your daily travel, i.e., walking to and from work and walking to shops instead of taking your car.

Increase you duration and start to walk over the countryside and hilly terrain. Build on this until you are walking some 20 km (10 mi) over hilly or mountainous terrain at least twice a week.

Running

Running or jogging is nothing more than fast walking, and here I give you a warning. While you will need the speed and fitness, you also need to be able to carry a heavy backpack and a rifle.

Start your running program only after two weeks of walking, especially if you are unfit. Once you start running, concentrate on a pace that suits you. Run with a heel-first action, letting the toes claw into the pace and pushing off with the ball of your foot. Run within your breathing capacities.

Week 1

- Warm up. Walk 3 mi in the morning and 3 mi in the late afternoon or early evening. Choose a flat route along the side of a river, etc., avoid roads if possible. Do this each day; skip Saturday, and walk one route of 6 mi on Sunday. Time: 1 hr.

Week 2

- Warm up. Increase your walk to 4 mi in the morning and 4 mi in the evening. Concentrate on speed, and carry a small backpack 10 kg (22 lb) max. Skip Saturday, and do an 8-mile walk on Sunday, again carrying a small backpack. Time: 1 hr 15 min.

Week 3

- Day 1: Once only, walk out 5 mi with backpack—run back. Time: 1 hr 30 min.
- Day 3: Once only, walk out 6 mi with backpack—run back. Time: 1 hr 45 min.
- Day 5: Once only, run out 6 mi with backpack—walk back. Time: 1 hr 45 min.
- Sunday: Run at a steady pace with backpack for 12 mi. Time: 1 hr 35 min.

Week 4

- Day 1: Run with backpack 6 mi—walk back.
  Time: 1 hr 45 min.
- Day 2: Sport of your choice: swimming, football, etc., for 1 hr.
- Day 3: Run with backpack 8 mi—walk back.
  Time: 2 hr 20 min.
- Day 4: Sport of your choice: swimming, football, etc., for 1 hr.
- Day 5: Walk and run for 2 hr with backpack; time your mileage average.
- Sunday: Select a hill walk of some 15 mi; increase the backpack weight to 15 kg (33 lb).

After four weeks, you should be ready to tackle rugged terrain on a more regular basis. If you are not always able to walk in the hills and mountains, then continue with week four, making your own variations. Once you have overcome the first month, your training will become easier and less boring. You will then want to push yourself and really get fit.

Food

Food is the fuel that feeds our bodies and keeps us going. It provides the building blocks for growth and repair. What you eat will have a direct effect on how your body performs, on your health, and on how long you live. In the normal course of events, people watch what food they eat to control their weight and achieve an attractive figure. Well forget about counting the calories. When doing strenuous military training, eat whatever you want. However, it is advisable to eat a mixture of foods so that your body gets an even supply of proteins, fats, and carbohydrates.

Eating is one of the more pleasant daily functions of life. During your employment, eat and enjoy all the food you can cram into your stomach; 2,500 calories are the normal intake for an active man, but as an active soldier of fortune, you should plan to eat at least 4,000–5,000 calories a day.

The good, the bad, and the ugly.

The Good, the Bad, and the Ugly

So you still want to be a gun for hire? Then this is what you need to know, and you must also be good at it. Even in "controlled" wars, there is one hell of a lot of ordnance flying around as well as gunships and drones that can see and kill you from miles away. . . . And if you are driving your Datsun Cherry dressed in civilian clothing with an AK-47 on your lap and you suddenly pull up behind an American convoy, you are very likely to get wasted in a friendly fire incident. However, your immediate concern will be some young trigger-happy kid or religious maniac who is just waiting to shoot someone, or the hidden Improvised explosive device (IED) just along the track. Then there is the threat of kidnapping: You can easily get taken hostage by some unknown radical group, placed in some undetectable location

Barry Davies

Stay down when under fire and you will stay alive.

where you are beaten and abused, and if no one pays the ransom, they will remove your head. My advice when you enter theater is to make sure you are wearing the right protection—head to toe—and are clued up and tractable. Above all, just know what you're getting into. But before you start thinking about all the money, you need to start thinking about surviving long enough to spend it.

Author's Note: If you have never been under fire before, when the tiny copper-coated bees start singing around your ears, hit the dirt. Getting yourself behind good cover is best, but even lying flat on the ground is better than standing up. Rounds that come close to you will make a "zipping" sound whereas the ones that are very close will crack. You will not hear the one that kills you. The first time you come under fire, there will be a tendency to shoot back in just about any direction. Resist the urge to pull the trigger, unless you have a definite target and that target is a positive threat. Never waste ammo: You never know when you are going to need it.

The Job

Before you go off and join the nearest available PMC, first consider what you are good at. For example, PMCs do not just want ex-soldiers; they also need pilots, engineers, experts in water drilling, and a whole variety of non-military skills. Many of these skills are required in danger zones and are extremely well paid, plus the PMC will normally provide a degree of protection. So firstly, when looking for a job, see what skills are needed and where, and go for the companies that offer the best pay and security. If you are ex-military, you will most likely know some old colleagues who are already working for a PMC; go and talk to them as they may have inside knowledge about which PMC or other is currently recruiting. Trust me, many jobs are found this way. The important thing to remember is to do your homework and not to rush into the first job that becomes available.

There are several agencies that deal with recruiting for the PMCs. If you cannot immediately find work, it is at least worth putting your CV on file with one of them. Also, search the web for PMC sites, as there are a lot out there and many will happily take a look at your CV. Research into which PMCs have been awarded a contract as if it has just happened, it is very likely that they will be looking for specialists. Finally, keep an eye on the TV and see where the wars and conflicts are taking place, what is happening in the conflict zone, and most importantly, when they start to rebuild the infrastructure.

Interview

If you get an interview for work, make sure you are ready to answer some tough questions. Relax and be yourself, do not go in all "gung ho" pretending to be Rambo reincarnated; you will not last two minutes. Before you go for an interview with any PMC, make sure you have as much information at your fingertips with regards to your abilities and security clearance. If in the first instance you are asked to send in your CV, make sure it is as detailed as possible. Many private military firms require extreme levels of security clearance and a relatively clear background picture of your history. You will also need to furnish at least two character witnesses: Choose former commanders with a good rank, such as a General you may have served under, etc.

During the interview, keep your mouth shut until asked a question, and later ask only simple questions about the work. Tell the truth. If this is your first interview, promote your skills heavily, but do not promise anything you cannot deliver. When it is your turn to ask questions, inquire as to which country you will be working in (if you do not know already, and if you do you need to research it) and where within the country you will be placed. If the location sounds isolated, ask more about it and also how many others will be working there. What exactly does the job entail? Ask about security, whether you will need to be armed, and what security is provided, if any. Once satisfied, ask about the more routine things, such as work/home furlough rotation, accommodation, food and transport, etc. DO NOT ASK ABOUT MONEY at this stage.

Once you are contented that you are up to doing the work and that the country and location are fairly secure, only then should you enquire about the contract. Ask what terms are they offering, what insurance they provide, etc. Now you can ask about the money. Then when you have a full understanding about the entire employment, make up your mind. If the job and the money sound too good to be true, then they most probably are: Check out everything. Make sure you know where you are going, what you will be doing, and who you are actually working for, especially if it's a sub-contract. If this is the case, ask which company is prime, so you can do a little research.

There are many good PMCs currently operating around the world; some you will already know due to their very size, but many wish to remain anonymous. To aid you in finding a PMC that could require your skills, you will need to sift through the list in Annex A. I have chosen a few of the more well known to highlight the diversity among them.

Blackwater and Other Notable PMCs

Blackwater, the name of which derives from the color of the swamp water of their training area, is a U.S. company formed in 1997 by Erik Prince, a former US Navy SEAL. The concept of the company was to provide both military and law enforcement agencies with good training. Due to the location and facilities of the training camp that opened officially in May 1998, Blackwater picked up a fair amount of training for the SEALs and various SWAT teams. However, their first real government contract came shortly after the attack on the USS *Cole* in Aden, Yemen.

Blackwater went from strength to strength and, due to its professional approach was awarded contracts with the DoD and the CIA, where it undertook top-secret work. After Hurricane Katrina, many other private companies also became clients of Blackwater, including insurance, petrochemical, and communication companies.

At first all was well with Blackwater, which had earned an excellent reputation both for its hard work and the good personal discipline of its operators. Then in September 2007, a shooting incident in Nisour Square, Baghdad, put the contractor into the spotlight in a very unfortunate way. At that time, Blackwater was providing security for U.S. diplomats in Iraq. A convoy carrying diplomats was approaching the square when a second Blackwater convoy, positioned on the square in advance to control traffic, opened fire, killing seventeen people and wounding another twenty-seven. It has been claimed that the occupants of the Blackwater vehicle opened fire at innocent civilians around Nisour Square without provocation, but the Blackwater operators involved claimed they had come under attack and had returned fire in self-defense. The Iraqis have disputed this, saying that not even a single brick was thrown. However, the law gives American contractors virtual immunity from prosecution in Iraq, and those responsible were free to return to the United States, angering the Iraqi authorities.

The author some twenty years ago (second from the left) training the Saudi Royal Guard in counter-terrorism.

Barry Davies

Blackwater drew a lot of criticism, not just for the killings, but also for alleged corruption and sanction busting. In addition, Blackwater's own records in 2007 indicate that company operators opened fire 195 times that year and fired first in 163 incidents.

Despite all this, Blackwater survived, mainly due to the larger world in which it operates and the professional leadership of the company. It has had many names, especially recently: Blackwater Worldwide, Xe Services LLC, and is currently called Academi. Academi is the largest of the private security contractors employed by the Obama administration, supplying diplomatic services as well as working for the U.S. Department of State and the CIA. The current contract is estimated to be around $250 million.

* * *

Of all the PMCs, the one that I knew the most about and respected more than any of the others was Defense System Limited (DSL). This British firm had its office in London—not far from Buckingham Palace—and was originally run by Alistair Morrison, a retired SAS major. What was so unique about DSL? Well it kept itself low-key, had class, employed the best operators, and pulled in some of the most lucrative contracts from the rich "blue chip" companies. It did what a PMC should do: operate with integrity and stay quietly in the background. Alistair Morrison went on to form Kroll some years later, once again a company well respected in the industry.

I also have a very good friend who is a former Royal Marine Commando (Scott Wilcox). Scott started his company with very little and today controls a multi-million dollar PMC called SicuroGroup. This company is a leader in expeditionary logistics and communications (which mean it provides tracking). Formed in November 2005, with its headquarters in Dubai, the company now has offices in London, Libya, Iraq, Afghanistan, and India. While the company will take on a variety of tasks, its specialty is to provide protective tracking of VIPs, media personnel, convoys, and even military vehicles. Scott uses state-of-the-art satellite tracking as well as local GSM and RF, thus providing a full range of options. If you are interested in SicuroGroup can use the contact page of its website, where you can leave your CV.[3] Do not bother to call them directly, unless you're a customer: Let them contact you.

Most PMCs will want to do a background check on the things you told them during your interview, unless you're from one of the specialist recruiting agencies. These checks will include drug screenings, criminal background screenings, and personal and professional reference checks. You will also be required to undergo rigorous physical-fitness screenings. Always tell the truth, as the past will always catch up with you, and do not bullshit: These people are professionals and will know.

If you have a past, you can bet your bottom dollar that the authorities will know about it . . . and they have a long memory. In late 2003, David Tomkins landed at Houston's George Bush International Airport. Officials had no reason to suspect that the polite sixty-three-year-old from the United Kingdom was any different from his fellow

---

[3] www.sicurogroup.com/contact-us.html.

passengers, until, that was, they scanned his passport. Seconds later, several armed guards pounced on Tomkins. You see, he had a record: He stood accused of masterminding a plot in 1991 to assassinate the Colombian drug baron, Pablo Escobar. More than a decade ago, Tomkins was caught in an undercover sting operation by U.S. Customs after he allegedly tried to buy a fighter jet to bomb a Colombian prison housing Escobar. According to the U.S. authorities, he had been on the run for the previous nine years and in his absence was pronounced as a "key player from the ranks of the international arms dealers." Tomkins was described as a man who had an "almost unquenchable thirst for adventure," and "any chance he had of doing something exciting, however dangerous, he took it."

Pablo Escobar, the notorious drug baron.

Author's Note: One of the most colorful characters I ever met, and a true soldier of fortune, was Pete McAleese. Pete started his life in the Parachute Regiment in 1960, and a year later volunteered for service with 22 SAS (22 SAS is the current designation of the regular British SAS). Being successful in selection, he joined D-Squadron and took part in operations in Aden and Borneo. He left the army in 1969, for life as a civilian . . . but obviously found no satisfaction in that, because in 1975 he was back on the battlefield. This time he was fighting as a mercenary in Angola, helping the FNLA (Le Front National de la Libération de l' Angola/ National Liberation Front of Angola) guerrillas in their war against the Cuban-backed MPLA (Movimento Popular de Libertação de Angola/People's Movement for the Liberation of Angola) government forces. His mercenary career lasted about a year, and he then returned home to England. It wasn't long, however, before Africa called him back, and he joined C-Squadron, Rhodesian SAS, serving with them until 1980. After Rhodesia fell into the hands of Robert Mugabe's regime, McAleese travelled to South Africa, joining the 44th Parachute Brigade. Achieving the rank of Warrant Officer, he was soon to see action in Angola once more, this time fighting against SWAPO (South West Africa People's Organization) guerrillas.

In 1986, he retired again, going back to England. Not a person to be settled in one place for long, McAleese again returned to Africa, this time working for a security company in Uganda. After that job ended, he found employment in Colombia, South America, where he was tasked with leading an attack on a Communist base in the jungle. This particular operation was called off when, in 1988, he was asked to carry out another operation by his employers: to assassinate Pablo Escobar, the leader of the Cali drug cartel. (It has never been established who his paymasters were at the time or if this mission were true.) Through his extensive contacts, McAleese formed up a force of twelve men and trained them for the mission. All seemed to be going well when disaster struck at the last minute. One of the helicopters carrying the men to the target crashed into a mountainside in mist, killing the pilot. The operation was aborted, and McAleese returned to Britain. (You might note that the previously mentioned Tomkins was also involved in the attempt to assassinate Pablo Escobar.)

At around the time McAleese was in Colombia, the British and U.S. governments were sending in Special Forces. The powers that be had decided that the SAS would help train the Colombian antinarcotics police. This would allow the Colombians to carry out their own special operations within a jungle environment. As a direct result of this training, one of the police units was able to track and kill Gonzalo Rodriguez Gacha (known in the drugs world as El Mexicano). There had been some speculation, although without foundation, that McAleese had been training Gacha's bodyguards. Whatever really happened, it has to be said that McAleese remains one of the most daring and adventurous soldiers of fortune ever.

Libya

I mention Libya because it is the latest place to find contract work. With the war now over and Gaddafi dead, the country needs to be rebuilt; but with the situation still volatile, the rebuilding contractors still need protection. Therefore, it is a very good time for the PMCs. They have been trying to move in since September 2011, and while some have made it and managed to get a contract, many are still waiting in the wings.

Prior to the conflict in Libya, there were approximately 80 to 100,000 contractors working in Libya: The Chinese had over 30,000, and the Indians had around 10,000. In addition, there were also many Egyptians, Americans, and Britons working there. Many worked in the oil fields which produced much of Libya's wealth, but others were employed on contracts ranging from building military installations to prisons, palaces, and anything else that Colonel Gaddafi thought he needed.

Once the uprising took hold and grew in size and strength, many of the regime's army sided with the protesters. Gaddafi then turned to the neighboring nations in order to recruit mercenaries to fight for him. These were mercenaries in the truest sense: young men who wanted to make easy money. Today, many of those frightened young men are sitting in jails, waiting for their new masters to determine their fate. The lesson here is not to find yourself on the losing side.

At the time of this writing, Libya's elections have been delayed by the National Transitional Council until July 7, 2012. This is important to the PMCs, as most contracts have been suspended until after the elections. Additionally, as the contracts are being awarded by the country itself and not an occupying force, they are not straightforward. For example, any company that had dealings with the Gaddafi regime will not be considered. Additionally, many of the Gulf States that are financially supporting Libya are having their say before contracts are finalized. That said, once the election is over, there will be some lucrative work coming out of Libya. Whether or not Western PMCs get a bite of the cherry will have to be seen, but one thing is for sure: Contractors will NOT get a jail-free clause for any misdemeanors, such as was seen in Iraq. (Latest news is a British firm has just landed a contract for $21 million to supply equipment for the new Libyan Boarder Regiment.)

Summary

As I have said throughout this chapter, being a soldier of fortune is not just about earning lots of money . . . it's a dangerous job; with high-paid work comes high risk. In a war zone, life is very unpredictable, and despite taking all the caution possible and developing a safety net, there are times when bad luck simply takes over.

Nick Berg was an American businessman from West Chester, Pennsylvania, and went to Iraq after the invasion. During his stay there, he was abducted, held captive, and then brutally beheaded in a video that was placed on the Internet for the world to see. Islamic militants who carried out the torture and beheading claimed that Berg's death was in retaliation for the atrocities carried out in Abu

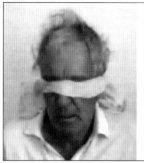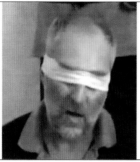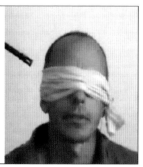

Left to right the men are Kenneth Bigleyard, Eugene Armstrong, and Jack Hensley; Eugene Armstrong was later decapitated.

Ghraib prison where U.S. service men and women allegedly sexually degraded and abused Iraqi prisoners.

Berg's body was found on May 8, 2004, on a Baghdad overpass by a passing U.S. military patrol. His family, who had already reported him missing, was informed of his death two days later. However, they were simply told that Berg's body had shown signs of trauma and at first were not informed that he had been decapitated. It was not until the film of the decapitation was released onto the Internet, reportedly hosted via Malaysia on an Islamic organization's webpage on May 11 that they learned the truth. In the video, Berg can be seen wearing an orange jumpsuit similar to that worn by the prisoners in Guantanamo Bay. He is surrounded by five men, all wearing ski masks. Prior to his death, a long statement is read out before the men converge on Berg and hold him down as another decapitates him to the continuous screaming of "Allahu Akbar" (Arabic for "God is Great").

Still, it would be wrong of me to impart that mercenary life is all ruthless and dangerous work; some people get lucky. A friend of mine—John Davies—joined the British SAS at the same time as I (around 1966), which puts him in his late sixties today, but his life story is one that would read from a comic book. In his younger years, John looked a bit like Garth, the blond-haired muscle-bound cartoon giant from the British newspaper the *Daily Mirror*. However, John was a quiet man even with his SAS comrades, and with girls came across as shy. This, coupled with the most appealing smile you have ever seen, won the hearts of many a lady. In the early 1970s, John found himself in the Oman war with the rest of his SAS Squadron. But as the war started to calm down, he became restless and started looking for adventure elsewhere. He found a job driving around the Saudi Arabian desert, recovering cars and trucks that had been abandoned. Then he went to work as a security guard at a British Embassy in Africa before finally finishing up as a bodyguard to a lady who owned half an oil company. His looks and quiet manner were not wasted on the lady, as he later married his Principal and to this day lives happily ever after, body guarding his wife.

2

WHAT A "SOLDIER OF FORTUNE" NEEDS TO KNOW

Working as a soldier of fortune is not just a matter of firing a weapon; it takes guts and some serious training to satisfy the masters of any PMC. If you have not served in the military before, you will need some serious skills and an understanding of modern tactics. The ability to carry out risk assessments, reconnaissance and surveillance, operation planning, and room combat techniques are only just a few of the skills that are required. Moreover, you will also need to be knowledgeable of enemy tactics in the theater of operation and how they would be used against you.

Terrorists can attack in any number of ways, and these are often savage and unforeseen. They may take hostages and hold them to ransom; alternatively, they may simply murder by bomb or bullet. It really depends on the situation and what the terrorist organization is trying to achieve. For example, if the attack is a reprisal, then it is almost certainly going to be a bombing or shooting incident.

To prevent such actions from taking place, the security forces monitor most of the indigenous terrorist organizations in order to anticipate any future actions they may be planning. As a soldier of fortune operating on behalf of a PMC, you will need to be appraised of the same information. Where the terrorists carry out a bomb or shooting attack, the security forces can only react retrospectively. Only when prior information is obtained can counter-measures be initiated to prevent or limit the damage caused by such an incident.

So what do you need to know? Well that is one big question, and the list is infinite and will depend very much on your job description. But in the main, if you are working for a PMC in a war zone, your skills should include some of the basics listed below:

- Risk assessment for an individual and for a business (e.g., oil and gas)
- Covert/overt and counter-surveillance techniques
- Advanced driving techniques
- Personal and residential escort duties
- Defense techniques against kidnapping
- Firearms training
- Close-quarter combat techniques (defensive and restraint)
- Basic first aid and advanced medic (see chapter 10)
- Executive travel protection
- Methods of instruction for training others

Risk Assessments

All war zones are high-risk areas, and it is important to gauge the level of violence and conflict to determine which areas are safe and which areas should be avoided. Some areas—while still war zones—may be less dangerous than others. Risk assessment is about companies or individuals who are preparing to visit or do business in these areas, as well as understanding the risks associated with each location. The decision to go into a war zone is down to the individual or company; whether or not a PMC is going to support and protect them will be down to the risk assessment. Be aware that some companies will take on really risky work whereas others will steer shy, no matter the financial incentive.

A. R. Howell

Risk assessment is all about analyzing the probability and taking action to negate possibility. This team is about to escort pipes from Kuwait into Iraq; their risk assessment will dictate the cost of transportation to the customer.

When a peace treaty has been negotiated to a safe standard, reconstruction will take place. This reconstruction takes many forms, from retraining the police and armed forces of the indigenous country to rebuilding roads, housing, waterworks, electricity, and just about everything that got damaged during the conflict. Contracts are issued by either the intervention country, such as the United States in Iraq, or the prevailing administration. (An example is the Transitional Council in Libya.) These contracts (and they can be very lucrative) will mostly be offered to civilian contractors (i.e., a construction firm if it involves building) or a PMC (if it involves retraining the police or military). In some cases, the civilian building contractor will require the protection of a PMC, especially if there is still a lot of hostility around. PMCs hired to provide protection services for other companies or the military should be involved with the contracting process from the beginning, with prime contractors building in a percentage to cover the cost of security. Some companies are ill prepared for this cost and have not built it into the contract during the bid process. As a result, the contractors are often unable to perform the work to their fullest capabilities. However, many contracts issued by governments will contain an "excusable delays clause," excusing the contractor from performance if "acts of God or of the public enemy" prohibit it.

So if you're going to do business of any kind in a war zone, the first thing to do is carry out a risk assessment to evaluate your cost model, allowing for insurance, healthcare, compensation, and generally, a significant chunk for security. In addition to military skills, there are also a few things you should also take into account, especially your personal insurance. Don't forget that most people have families who rely on them earning a good wage; the loss of the main wage earner could prove catastrophic to your loved ones. Before you leave your homeland, make sure you have good insurance coverage or are entitled to government assistance in the event of your death.

Insurance

One of the first things you will want to know about is insurance. As we have already seen, being a soldier of fortune carries a great deal of personal risk. While there are a few single men in the PSI, most have wives and children. It is imperative that prior to undertaking any work with a PMC, you check out what insurance security they provide.

Insurance underwriters are able to offer insurance for those PMCs using firearms, as well as the less risky work. However, always read the small print, as there is often a limit of indemnity clause restricting payouts to around the £1 million mark for death. Also, the premiums are very high.

In addition to their performance obligations, all U.S. government contractors and subcontractors working outside the United States must secure workers compensation insurance, known as Defense Base

Act (DBA) insurance, for their civilian employees, including U.S. citizens, third-country personnel, and local nationals. The contracting company must prove that it has purchased the mandatory insurance before it can be awarded the contract. DBA benefits cover the cost of medical treatment as a result of injuries received while a contractor is performing the job. It also reimburses wages lost because of injury and offers beneficiaries disability and death benefits.

Local Area Intelligence

There are two other things you will need before you depart: The first is to have a local area knowledge of where you're being sent. The second is a record of the PMC you are working for. The first can be gleaned from a quick search of the Internet, whereas knowledge of your PMC is best coming from others who have worked for it. Why do you need these? Well, your PMC might put you into a position that is highly dangerous; it might be short of cash and not pay you. It is always best to check out the track record of a PMC before signing any contract. Make sure the PMC does not cut corners to make an extra buck and that it supplies good protective equipment and vehicles.

Having the right people and proper equipment is important for the protection of any contractor in a war zone. However, it is intelligence that is the most important tool for keeping out of harm's way. You need to know what is going on in your area of operation. If you do not, then how can you provide any rational form of security? You will also need to establish a line of communication with any regular forces operating in the area; for example, if you are an American, then the U.S. military force is an obvious choice; "blue helmet" NATO forces are also a good source of local information. In many cases, occupying forces will hold regular situation updates as well as issue warnings of possible attacks. Local knowledge before traveling to avoid possible enemy hot spots or friendly fire will help keep you out of trouble. It is quite usual to have a lot of PMCs—large and small—working within the same area, and it is important for contractors to share intelligence information with each other. If the

Internet is available in your country of employment, then your desk office should keep up to date with all reported incidents. There are many sites which update on a daily basis; these will inform you not just of the latest attacks that have taken place but will provide you with an overall pattern of areas to avoid.

Barry Davies

Close-quarter battle begins with basic shooting skills.

Personal Skills

Even Special Forces members will start off learning their shooting skills at a basic level: pistol handling, concentrating on stance, accuracy, and simple Close-Quarter Battle (CQB) practices. At this stage, the basic moves—kneeling, turning, and rolling—are incorporated. At the same time, the practice of firing a "double tap," is taught (which means firing two shots in rapid succession). It takes several years to become comfortable with this method of shooting, but the results have proved themselves over and over again: Two rounds stop a terrorist better than one.

In the arena of PMCs, personal skills are all about being professional, staying alert, and doing your job to the best of your ability. Many people I've known have never fired a gun before or even held one prior to doing a civilian course; they then run off to the nearest war zone and get a job. Their drills tend to be sloppy, and while many

Barry Davies

Assault speed is everything. Picture shows two members of the British SAS assaulting after the wall has been breached.

never get into trouble, if they did have to protect themselves in a roadside ambush, they would most probably die. While I am limited in being able to impart much in the way of weapons skills in this book, I can highlight some of the more conventional skills you will need to learn, especially if you intend instructing others.

Assault Speed

In all my years, whether in defense or assault, I have learned one thing—speed. Much has been written about the swiftness of Special Forces during their assaults and with justifiable reason. These units learned many years ago that both speed and aggression play a major role in any successful assault or defense. The basic instinct of survival is strong in all humans, including those who are prepared to die for their cause. However, just prior to the "moment of truth," there is always a certain amount of hesitation. For example, while the terrorists are pointing guns at their hostages, threatening them with death, they are in control. However, the moment they are faced with figures all dressed in black and confronting them with imminent death, the situation changes. No matter how dedicated the terrorist, he will ignore the hostages and concentrate on the immediate threat to his own life. Unfortunately for the terrorist, the man in black has superior training, state-of-the-art weaponry, and body armor protection. There is at this stage no contest other than one of speed. Closing with the enemy is vital; the quicker it can be achieved, the greater are the chances of success. Moreover, any further risk to the hostages will be minimized.

It is the same for anyone working for a PMC in a war zone: The threat is constant, and you can find yourself being attacked at any time in any location. Your survival and that of those you may be protecting will depend on your assault speed. Speed can be achieved through rapid understanding of any sudden attack; a swift and deliberate defensive assault will help you recover the initiative. There must be minimum delay in the response time to any situation: You must react instantly. It is crucial to locate and physically close the distance to the point where any terrorist threat can be neutralized. Most retired soldiers know this and go into assault speed automatically; but if you are new to the business, learn to react very quickly.

Author's Note: While I am sure I have written this elsewhere in the book, it is worth repeating. If you are going to fire your weapon in defense or as a protective aggressive maneuver, aim for the head. In a war zone, an enemy with commitment to a religious ideal or running high on drugs is hard to stop, even when you have put several rounds into their body; put them down first time, and make sure they stay down with two in the head.

Barry Davies

Learn the basic skills, such as clearing a weapon jam quickly.

Basic Skills

The next level is to progress onto learning to use a pistol and a light machine gun. These weapons are like the third arm of a soldier and are used by just about every country in the world. What makes modern weapons so special are their compactness, power, and ability stop most enemies in their tracks. Special Forces will use a strap down pistol and normally carry a light machine gun. Many will shun weapon slings as being cumbersome; conversely, some machine guns operate better when secured against the body, allowing the side arm pistol to be used. You need to become very familiar with both your side arm and your machine gun: Know how to operate them, what causes any malfunction, and how to clear them quickly.

The final stage of weapons training is to use both a machine gun and pistol together. This incorporates practicing malfunction

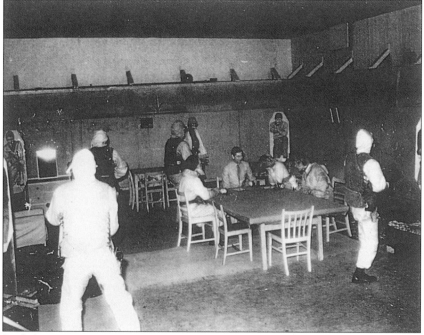

Barry Davies

Highly advanced skills were required as members of the British SAS rescued Prince Charles and Princess Diana in the killing house demonstration. Live ammunition was used.

drills on either of the weapons while using the other as an immediate back up. While all Special Forces soldiers learn malfunction drills, clearing a jam in the middle of an assault requires too much time, and therefore it is prudent to have a backup weapon. Once these basics are complete, students move on to a variety of hostage rescue scenarios, including the "snatch."

This is a drill that is practiced as part of an anti-terrorist hostage-rescue situation. It involves going into a room shrouded in darkness, in which a live hostage is to be found seated among several paper targets. The team must neutralize the "hostiles" and rescue the hostage. Practice is essential: On average it is unlikely that any member of the team would spend much less than 2,000 rounds of ammunition a week on one range or another. Modern shooting ranges and "Killing House" facilities are extremely sophisticated; but if you are not a member of the Special Forces, there are lots of places in America and Europe where you can go and get trained.

Author's Note: There is an endless procession of VIPs that visit the SAS in Hereford, and the best demonstration is to involve them in a "Snatch" scenario, using them as the hostage to be extracted. Many VIPs that visited Hereford have sat in the hot seat while black-clad figures burst into the darkened room firing live rounds at targets positioned within inches of the VIPs. The unsuspecting hostage will find himself sitting in absolute darkness, surrounded by silence, not knowing what possessed him to volunteer in the first place. Abruptly, the door bursts open, and a stun grenade explodes inches from him. Laser beams of light penetrate the blackness, searching hungrily for targets, followed by a hail of bullets. Luckily, VIPs are plucked, albeit roughly by the black-clad figures and literally thrown out of the room, surprised later to discover not a single scratch upon their precious persons.

If you have not been in the military and have no weapon handling experience, I would suggest you enroll in a private course immediately.

There are many excellent schools in both the United States and Europe where basic firing positions and combat techniques can be learned. I would recommend you pay particular attention to and, if possible, practice firing the AK-47. I say this for no other reason than it is the one weapon you are most likely to obtain in a war zone.

One of the major roles undertaken by PMCs is that of supplying bodyguards to visiting VIPs or other contractors. The word *bodyguard* has a lot of different definitions, but it basically means looking after others in your care. The role could be a simple matter of escorting a convoy of essential material from a safe location into the war zone, or it could be looking after a TV crew who want to get close to the action.

The prerequisite for a PMC CP team can be broadly divided into the following categories:

- Full CP team providing protection to a Principal at his orher residence/hotel and during his or her movements throughout the day. This will generally be a more high-profile operation, using back-up vehicles, communications, and enough personnel to provide a flexible protection plan throughout the twenty-four-hour period.

- A small, often two-man team to provide a client with a very discreet level of protection. This will generally involve one man remaining close, but not next to, the Principal and the other travelling ahead to arrange "meet and greet" at airports and to check accommodation or venues prior to arrival. Female operators can often provide discreet protection, particularly for female executives travelling overseas.

- A single man or woman—generally a very experienced operator—who travels with the Principal and audits the reliability and professionalism of a locally employed team. This is particularly common when local driving knowledge and interpreters are essential.

KIM Info Service

Close protection for the very Reverend Archimandrite Nektarios Serfes (Greek Orthodox Archdiocese of America) during his recent visit to Kosovo and Metohija.

PMCs do not simply offer protection, they also offer training. Again, this has a lot of different connotations. After a country has been devastated by conflict, law and order must be a top priority, or else wide-scale anarchy will prevail. (Mogadishu is a prime example.) A police force will need to be re-constructed, as will the army in many cases (as seen in Iraq and Afghanistan). Both will require training. In this case, governments will often employ a PMC to form a training team with the precise task of bringing some order back into a country.

In recent years, this has become a primary role of many PMCs: training the police, military, or even a country's Special Forces (the latter is normally done country to country). One of the prime skill sets in training is teaching counter-terrorism methods. This subject has been defined and refined to the point where it is almost a set pattern worldwide with slight variations depending on the threat.

Barry Davies

Training teams start with the basics. Here students are being instructed in methods of entry using explosives.

Training Team

If you get offered a job on a training team and you feel your skills are up to it, then you should give it a go. Training other personnel is a good job and is mainly carried out in a safe location. However, before any training can begin, you will need to know who you are training and where. The "who" will dictate what standard they are at, and thus at what point you will need to start the training; the "where" will dictate the type of training facilities you will have to work with.

The type of training to be carried out and the number of people to be trained will dictate how large the training team should be. For example, training a thirty-man team in counter-terrorist techniques over a period of six weeks will require a training team of eight men. A preliminary recon (recce) will be carried out by a senior member of the PMC prior to the main team's arrival. This recon will establish all the requirements needed and help formulate the number of trainers required and what skills they should teach; at the same time, accommodation, training and classroom facilities, ranges, and transport will also be discussed. Of prime concern will be security of the housing and primary training area: They must be discreet, secure, yet offer lots of buildings of various types and have suitable vehicles and roads.

A training team will normally start off by doing a discreet assessment of the manpower they have to train to ascertain their standard. The basics of weapon training and handling is a good place to start. The course will then progress, working up to full practice attacks on both buildings and aircraft. In countries where the climate and religion are a consideration, the daily program will need to be adjusted to accommodate both the mid-day heat and the religious prayer. For example, morning classes might normally start with physical exercises at 5:00 a.m. followed by classroom lessons. Breakfast is then taken around 9:00 a.m., with lunch at 1:00 p.m. A sleep break then follows until lessons resume between 5:00 and 7:00 p.m.

Classroom lessons cover dry weapons training, including pistol and automatic weapons. On week two, range work will begin. If none are available, special ranges will need to be constructed, both to cover normal CQB shooting and a special range where "Killing House" techniques can be practiced. In the case of the latter, the walls of the building are normally constructed out of hessian cloth and secured behind a sand bank. Range safety is paramount, and thus command and control on the ranges requires the attendance of several team members at one time. The students should progress from basic pistol work and firing double taps to short controlled bursts with the automatic weapons. As the students move on to the makeshift killing house, they will form into teams, learning the more difficult house-clearing techniques. In addition to buildings, aircraft, ship, and rail assaults may also be required and practiced. All of this will lead up to a final exercise in which everything they have been taught can be demonstrated.

Barry Davies

This training team were working in Africa. Here they can be seen attacking an enemy encampment supported by a helicopter.

In order to teach counter-terrorist skills, the instructors must first have an in-depth understanding of each and every subject, and this is not something one can learn from a book—it must be knowledge gained from direct experience. It begins with understanding what a counter-terrorist unit is, how it is formed, and what purpose it has. Also how each team member plays his or her part, including his or her role and responsibilities within the larger scope of the operation. Instruction will start with the basics, with an assessment of the students and their current standards in shooting and tactics. The counter-terrorist course will progress with the students learning skills such as entry techniques, room combat, how to enter an aircraft or ship, and how to deal with the terrorists. A counter-terrorist course is normally done in phases, with each phase being practiced through a small scenario, finally building into a full-scale exercise.

Ideally, training should be carried out in a facility that specializes in counter-terrorist training; unfortunately these are few and far between. Most Special Forces will have their own training facilities, which include short and long firing ranges, a range of buildings for assaults, an aircraft, a mocked-up ship (both at sea and alongside), a mock airport, train and coaches on rail track, a stretch of motorway, and many others. Military facilities are generally for military use only, but there are both government and private training facilities available in many countries. Access to these facilities will very much depend on the reputation of the PMC and who it is training, i.e., are the students from a nation that is friendly to the West?

Building Assaults

All building assaults pose the same problems: the number of floors involved and the number and distribution of hostages, together with the number and location of terrorists. Getting close to a building is in most cases fairly easy, as is gaining access. The major problem rests with closing with the terrorists quickly enough to prevent them from killing any hostages. The best-known building assault was that

Barry Davies

More advanced training will take you onto building assaults and how to gain accesses at different levels using an intervention vehicle.

of the Iranian Embassy in London, and this lends itself to being a good example of building assault. Any building assault will require knowledge of diversionary techniques, methods of entry, and room-clearance techniques. The range of buildings that may come under terrorist control and therefore need such an assault is very wide and can be anything from a skyscraper to a single one-story building. But the skills that must be learned should cover every eventuality.

Aircraft Assaults

Having carried out an aircraft assault personally, I think I can safely offer some advice. Hijacking means that terrorists have taken over an aircraft, where in the past this was to have demands met for the freedom of the passengers. More recently, however, they have a nasty habit of crashing into buildings.

Barry Davies

Aircraft assaults are not easy; you need to know all the blind spots both for approach and when gaining access into the aircraft.

To assault any aircraft, it must be on the ground (don't believe what you see in movies about air to air recues . . . never going to happen). The aircraft needs to be identified and, if possible, isolated from the general public. You will need to know the internal layout, the number of passenger, crew, and hijackers if possible. The main points for consideration when planning an anti-hijack are as follows:

- Type of aircraft and airline, as internal configurations differ slightly.
- Location of aircraft for your approach.
- Number of passengers and crew to assess how many seats are filled.
- Number of terrorists and weapons/explosives.

- Method of entry: How would you best gain access into the aircraft?
- Assault pattern: Usually multiple entry at the same time.

Once the team has reached the hijacked aircraft, it needs to establish full data and activates of the hijacked aircraft. If not already done, insert sniper teams covering all aspects of the aircraft. Develop an immediate action plan just in case the terrorists actually start shooting hostages, i.e., ready the intervention vehicles. Carry out a CTR (Close Target Recon) and route to aircraft, both overt and covert. Develop a firm assault and distraction plan.

Almost all aircrafts are accessed by either hand-carried ladders or platform vehicles. Once the type of aircraft and the seating configuration are known, the basic assault ladders and assault vehicles can be assembled accordingly. Method of entry will depend on the assault plan; however, the team commander has a number of options. Experience has shown that using the passenger service doors is the best option, as these can be operated quickly from outside the aircraft. Entry using the passenger service doors normally involves opening the over-wing escape/emergency hatches as a secondary method of entry. Window entry is possible, but this is size restrictive, and the flight deck is almost always occupied by the hijackers. However, it is useful for defeating individual hijackers who occupy the flight deck. While access can be made through the underbelly and lower freight doors, it is time consuming and noisy.

Finally, there is the option to chop out sections of the aircraft in order to gain access to the passenger deck. This is highly specialized work that requires skillful cutting techniques. While the safety of the passengers is paramount, cutting into an aircraft using thermal cutters or explosives will almost certainly render it inoperable for further use. This form of entry is also a risk to the passengers' safety.

Once access into the aircraft has been gained, the idea is to close with the terrorists at maximum speed and eliminate them quickly, then remove the passengers off the aircraft and assemble them at the rear ready for clearance transport, under control, i.e., no headless chickens.

Authors Note: It has been my experience that once inside the aircraft and closing with the terrorists, in the last few seconds of their life, they seem to forget about killing the hostages and try to stop the assault team members bearing down on them—at which time it's all over.

Ships and Rigs

In the unusual case of a ship which is tied up alongside a dock being hijacked, it will normally be treated in a similar fashion to a building assault. However, this scenario is very rare, and most terrorist threats against shipping have taken place at sea. As with oil rigs, this threat poses several problems, especially when approaching the ship. A ship or an oil rig is isolated on a flat surface, over which hostile forces can observe for a distance of at least several miles, or much greater if a radar is used. This limits the element of surprise and thus gives the terrorists time to carry out retaliatory actions prior to any assault.

In the past, anti-terrorist team members have been dropped by parachute close to the ship, but this method of deployment is very hit-and-miss and equally dependent on the weather. A sub-surface assault offers the best approach. This can be achieved by the use of submarines, sub-skimmers, or divers—all which offer excellent covert delivery to the target. The problem of getting from the sea surface and onto the ship or rig superstructure also has to be overcome. Although oil rigs are tall structures, most offer some form of ladder system that will allow access to the main superstructure. Smaller ships that are not too high above water level can be reached using a special grapple and flexible ladder, which is hoisted into position. However, larger ships, with a tall superstructure will need the deployment of high-suction pads to climb the outer surface. This can even be done while the ship is moving. Once the assaulting team has several members on board, they will then release flexible ladders over the side to enable other team members to climb. The actual assault is carried out in a similar way to a building assault, with each deck being cleared from the top down.

The biggest threat against both shipping and oil rigs is one of total destruction. A small motor-powered launch fitted with several tons of high explosive would be enough to send either to the bottom of the sea, and as yet there is no answer to this potential type of terrorist attack.

Barry Davies

Controlling a bus assault means stopping it in exactly the right place and gaining entry to overpower the terrorists—all in a few seconds.

Trains and Buses

Trains and buses are very similar to assault in as much as both are capable of movement along a linear axis, i.e., a road or rail line. Equally, they both require stopping before any assault can be made.

With buses, it is possible (if the bus is supplied by the assaulting force) to attach a device capable of stopping the vehicle at any given point. Trains can be stopped in a similar way, but not with the same degree of accuracy. To overcome these problems, the SAS worked on a number of assault methods, including dropping men from helicopters onto trains and buses. Additionally, both methods of transport are limited to speed and destination, which means that the anti-terrorist team can at least keep up with the hijacked transport, assaulting it when the opportunity arises.

Barry Davies

It is also possible to use a helicopter to assault a moving vehicle such as a bus.

Once stopped, both train and bus can be entered in a similar manner. Traditionally, this is simply done by placing ladders against the vehicle and smashing in the windows and doors. As both means of transportation are fairly narrow, it is not always necessary to enter the bus into order to pacify any terrorists. If there is a need to enter, then stun grenades and CS gas provide an excellent means of distraction.

Distractions are simply a way of taking the terrorists' minds off killing someone once they hear or see any assault about to take place. It gives the assault team a few vital seconds with which to close on the enemy. Explosive charges, while used for entry, can also be a great distraction.

Barry Davies

Sometimes you just have to blow a hole in the wall in order to confuse the terrorists and gain entry.

Author's Note: Training others in counter-terrorist techniques can be a double-edged sword. In many cases you may be teaching potential terrorists all they need to know about modern CT methods. Therefore, before any training can take place, several things must be in place. First and foremost, does the PMC have a contract or permission to provide the training from their government? If not, is there any reason why they should not carry out the requested training? This will be followed by a detailed synopsis being provided to the end user (normally a government or recognized organization), and a price agreed upon acceptance. A typical synopsis and offer letter is set out below. This is an old document I wrote for a government CT training proposal. I have changed the names to "Somewhere," and for the PMC, "Training International Limited." This name is totally fictitious and bears no attachment or association with any known PMC of that name. Readers are free to use the outline proposal below and modify it to suit their own training program.

3

THE ENEMY

The reason you are working for a PMC is that they have been contracted to provide protection in a dangerous area or war zone. In doing so, you will need to know what you are up against and who your enemies are. You will need to know how they operate, where they are most active, and what risk do they represent. Once again, I cannot help you with this, as there are so many violent factions operating around the world; you need to do your own research. However, I can offer you some general advice on terrorist groups and what they are fighting for.

One of the major differences in any social structure is the variety of religious beliefs, which have formed into blocks around the world. Some of these blocks share many common elements of everyday life, in particular religion beliefs such as in the Middle East. It is within these social groups that we find the answer to one of the root causes of terrorism.

For example, ask me what I believe in, and I will tell you it is my fellow man. I have lived a good life and fought in many wars where the background hatred has been religiously motivated. I am not a

Muslim, Hindu, Buddhist, or Catholic; therefore, I cannot say how the people of those religions feel towards their fellow man. I am neither a practicing Christian. True, I grew up in a Christian society and attend such Christian activities as weddings and funerals, but in my heart I have seen many issues with religion and the deaths that it has caused. It is also wrong to think that all religious hatred derives from the Muslim world, as Catholics and Protestants have been warring for centuries. Even when the same God has been on both sides, both warring parties have claimed that they were right and their enemy was wrong. Yet, despite all this death and pain, I have seen so much good come out of the human race. The struggle of one human to save the life of another knows no bounds, and therefore, I find that most humans are basically good. Where humans fail, they fail through fear and the generation of fear.

The attack on the twin towers in New York City must rank as one of the deadliest terrorists episodes of all time.

Many people have tried to establish a definition of terrorism, and while all are correct, none actually expresses how terrorist acts develop. The United States, in particular, is keen on establishing a firm definition for terrorism (although this may be because they wish to find a legal definition to justify retaliation). World terrorism is so diverse that to try to pin a single label on it is dangerous, although many have tried.

> "Terrorism constitutes the illegitimate use of force to achieve a political objective when innocent people are targeted."
>
> —Walter Laqueur
>
> "Terrorism . . . any type of political violence that lacks an adequate moral or legal justification, regardless of whether the actor is a revolutionary group or a government."
>
> —Richard A. Falk
>
> "Terrorism can be briefly defined as the systematic use of murder, injury, and destruction, or threat of same, to create a climate of terror, to publicize a cause, and to intimidate a wider target into conceding to the terrorists' aims."
>
> —Paul Wilkinson
>
> "Terrorism is the use or threatened use of force designed to bring about political change."
>
> —Brian Jenkins

In a similar vein to that of the U.S. Department of State, Paul Pillar, former deputy chief of the CIA Counterterrorist Center, defined the four key elements of terrorism:

- It is premeditated—planned in advance, rather than an impulsive act of rage.
- It is political—not criminal, like the violence that groups such as the mafia use to get money, but designed to change the existing political order.
- It is aimed at civilians—not at military targets or combat-ready troops.

- It is carried out by sub-national groups—not by the army of a country.

Again, much of this is correct, but the same could be said about World War II. The question still goes unanswered: Why is it premeditated, why is it political, and finally, why do innocent civilians suffer more than the military?

In reality, we must all arrive at our own definition of terrorism. We cannot place our own definition on others of differing territorial, social, or religious backgrounds. Where there is to be a definition for terrorism, it must come from the majority and not from the powerful.

The reality, is simple. Despite what you might think, very few of us are born free. Our parents and our immediate society will be governed by a set of rules and laws. From the moment we enter the world, our parents start the process of teaching us these rules in an effort to guide our lives as we mature through life. In addition to these, we also learn basic social habits from our peers, relatives, and school friends. By the time we reach our late teens, we know where we fit within our own society and are ready to face the world. However, when we venture into the world, we are confronted by different types of societies, some of which are intimidating or even hostile. This reaction is a direct result of our earlier social education. That is to say, the sum of our national pride, religious belief, prosperity or poverty, and the political values of the family unit all form preconceptions in our mind towards those of other ethnic groups.

History and many wars have shown us that this divergence of ethnic groups has provided the human race with untold permutations for conflict, denying any definitive center line for terrorism other than death and destruction. Therefore, we could argue that world terrorism is a collection of differing political, social, and religious beliefs, of which we are all a part. From the moment we are born, we become involved; it is simply the order of things.

Terrorism only exists within the human social framework, and individuals within that framework define and execute terrorist acts. In order to prevent terrorism, we must first understand the human element. For example, while America wept on September 11, many Muslims danced with joy. This role was reversed when Osama bin Laden was killed: Americans were happy while many of his followers threatened revenge. The value that individuals choose to accept in their own community allows them to judge what is morally right when an act of violence is committed. While these judgments may differ radically from one community to another, in the final analysis, it is an alliance of those communities that share a common definition in their beliefs that is most likely to succeed in winning the struggle.

In order to fully understand the individual's point of view, we should compare the framework that eventually defines individuals in relation to terrorism. As previously stated, people do not just become terrorists or anti-hijack experts; there is always a basic cornerstone on which all lives turn.

Author's Note: Having worked in just about every corner of this planet, I have come to one very clear conclusion: There are those who have and those who have not—money is the root of all evil. I have often thought that many of the conflict zones could have avoided confrontation had those in charge provided for their people better than they did. Much of the world's wealth is held and controlled by a few; therefore, the majority of people struggle to get the daily needs to sustain their families. Even when there has been a revolution, many leaders forsake their basic principles in the face of newfound power and riches. It takes very little to make impoverished people happy: All they need is a job and bread on the table to feed their children. Take away these basics and you have insurrection.

Lise Aseru`d

Religious ideology. The Norwegian anti-Muslim fanatic Anders Behring Breivik looks on during the morning break on the sixth day of his trial in Oslo, April 23, 2012.

Religious Ideology

Perhaps religious ideology is the wrong word to use, but I cannot understand why some people believe that their religion is superior to another. In the West, we tend to see the Islamic militant world as the enemy, but this is not so. I have visited many Islamic countries, and to be honest, their societies are little different from our own. Brunei, Malaysia, and Oman are just three examples of Muslim societies where I could happily live out the rest of my days. In the end it is down to the individual state and its people, not the religion. An example to the contrary argument is told in the story below, which took place in Israel in 1994.

Author's Note: Born into a middle-class neighborhood in Brooklyn, New York, Benjamin Goldstein, or Benjy as he was known, was a well-liked religious boy. The young Goldstein, with his traditional side curls and yarmulke, attended school at a yeshiva where his faith seemed to draw him into isolation, parting him from others. Goldstein went on to gain a medical degree at the Albert Einstein College of Medicine, after which he immigrated to Israel in 1983. Soon after, he married and eventually settled in Kiryat Arba, a West Bank settlement just outside the Palestinian city of Hebron.

He worked as an army doctor at the local clinic, but Goldstein was something of an enigma within his profession, as he would not treat Arabs. In 1990, when his hero, Rabbi Kahane, was shot dead by an Arab assassin in New York City, Goldstein's hatred deepened, and he vowed revenge for Kahane's death.

On Friday, February 25, 1994, the Islamic holy day during Ramadan, the month of fasting and prayer, Goldstein walked to the Tomb of the Patriarchs. It was the Muslims' time for prayer, and Ramadan guaranteed a wall-to-wall crowd of worshippers. By 5:20 a.m., about 700 men, women, and children jammed into the mosque for the dawn prayers. A mosque guard saw Goldstein approach and, recognizing him as a troublemaker, forbade him to enter. In reply, Goldstein, wearing the olive-green uniform of an army reserve captain, unslung the military-issue Galil assault rifle and swung the butt at the guard, knocking him down before running into the mosque. Inside, Goldstein took up position facing the backs of the worshippers, who knelt row upon row, heads bent in prayer. From less than 2 m (2.5 ft) away, he opened fire.

The bullets tore into the wall of humanity, hitting the people in the back and head, the heavy Galil automatic dispensing death. Unmoved by the total carnage about him, Goldstein changed clip after clip. Palestinians were running in an effort to escape; all the while the bodies of the wounded and dead mounted up in a sea of blood, the confusion making it difficult to distinguish who still lived among them.

The firing inside the mosque continued for about ten minutes until at last some of the worshippers managed to hit Goldstein with a fire extinguisher, momentarily disarming him. Three Israeli soldiers who entered the mosque at this time interpreted the scene as an attack by Palestinians on a uniformed Israeli and opened fire, killing several more. Outside, several other Israeli soldiers, who had just arrived on the scene, simply panicked as hundreds of frantic and bleeding Palestinians fled the slaughter, and they too opened fire. The final tally of dead and wounded exceeded even the number Goldstein could have wished for: Israeli officials counted thirty-nine people killed at the mosque, although the Palestinians figured fifty-two, plus seventy wounded. These figures did not include that of Goldstein, whose body was beaten to a pulp.

As the news spread throughout the fanatic Jewish precincts in Hebron, settlers danced in the streets and praised Goldstein's martyrdom, and some residents of Kiryat Arba called his act "a great gift." One settler, stopped by a soldier as she tried to assault a Palestinian journalist, shrieked, "We should kill 500, not 50!"

Later, as the bodies from the carnage in the mosque were jammed into ambulances without pausing to sort the living from the dead, ambulance driver Khaled Jaabry discovered, when he reached the local hospital, that among the wounded he was carrying were his own son and brother.

Al-Qaeda

This terrorist group has become the greatest potential threat against the national security of many countries, especially after the World Trade Center atrocity on September 11, 2001. With bases and supporters in many countries, it embraces and supports radical Islamic

Al-Qaeda rebels have been fighting for years, and despite their inferior weapons, their tactics are really good.

movements worldwide, urging them to overthrow "heretic" and pro-Western Muslim governments and to bring down the United States, who it sees as its prime enemy.

The organization was formed in 1988 by Osama bin Laden, a radical Islamist from a wealthy Saudi Arabian family. Bin Laden was one of fifty-three children of Saudi construction magnate Muhammad Awad bin Laden. His mother was reportedly a Palestinian, and the least favored of his father's ten wives. The elder bin Laden moved to Saudi Arabia from Yemen and amassed a fortune. Most of this money came from a number of successful construction and contracting companies. Today, the bin Laden family fortune is estimated at $5 billion.

Bin Laden entered on the path of holy warrior in 1979, the year Soviet troops invaded Afghanistan. Recognizing the Afghans were lacking both infrastructure and manpower to fight a prolonged conflict, he set about resolving both problems at once. Using his wealth, bin Laden organized conscription and advertised all over the Arab world for young Muslims to come fight in Afghanistan. He paid for their passage and set up facilities to train them. Bin Laden brought in experts from all over the world on guerrilla warfare, sabotage, and covert operations. Within less than a year, he had thousands of volunteers in training camps.

His army formed a major part of the resistance movement against the Soviet invaders. He saw that the Muslim fighters were well equipped and could carry out a sustained conflict against a technically superior enemy. Together with the Palestinian Muslim Brotherhood leader Abdallah Azzam, he began to organize an international program of conscription through a newly formed recruiting office: Maktab al-Khidamet (MAK Services Office). Ironically, at this time, bin Laden was probably aided by the Western powers in his struggle. NATO, and in particular the United States, saw Russia's invasion as a threat to its own presence in the area. The CIA launched a multimillion-dollar operation to train and equip the rebels—or Mujahedin, as they became known—and this may also have included some of bin Laden's men. With up-to-date weaponry, training and discipline, the Mujahedin eventually proved more than a match for the Soviet troops, and the invading army was forced into an ignominious retreat ten years later.

By the end of the war, bin Laden had around 10,000 well-trained, equipped, and war-hardened veterans under his command. Some dispersed back to their former lives and families, deciding that they had done their duty and could now live in peace. Others had become so fanatical about Islamic fundamentalism that they returned to their own countries, determined to organize their own terrorist groups and overthrow what they saw as "secular" governments. Egypt and Algeria both saw the rise of dangerous new extremist groups intent on causing death and injury to those whom they saw as obstacles in their way to a new fundamentalist country. Other countries that already had a leadership based on Shariah law welcomed the Mujahedin and also Osama bin Laden. Sudan, in particular, was friendly towards him, providing jobs and training facilities for his men.

Barry Davies

Terrorist training camps can be found mainly in the Middle East. This one is in the deserts of West Africa. Rebels training in room combat techniques.

The end of the Afghan-Soviet war was, in many ways, a watershed for the many Islamic groups who had been formed for the purpose of resistance. In 1988, bin Laden split from his MAK co-founder, Abdallah Azzam, in order to form a group that was to carry on his vision for radical Islamic superiority. That group was called al-Qaeda. The following year, Azzam was murdered by a car bomb. Despite there being a number of likely suspects, a rumor still lingers that bin Laden was behind it. However, as with a number of acts of terrorism linked to his name, it has never been proven.

Bin Laden's new fight started with his own country of origin. After the Afghan war, he returned to Saudi Arabia to raise an insurrection against the government. However, this plan was thwarted, and bin Laden was thrown out of the country and his citizenship revoked. Undaunted, he moved his center of operations to Sudan, using his wealth to create farms and factories while being able to supply jobs for his followers. He also improved the infrastructure of the county—building roads and an airport. His many business interests included a bank, a construction company, and an import-export business. More training camps were set up, and former Afghan comrades were encouraged to move to Sudan.

However, Sudan's friendliness towards its guests soon began to crumble when its government started to bow to pressure from the United States. Not wishing to be subject to sanctions or other methods of persuasion, the Sudanese asked bin Laden to leave. This he did in 1996, returning to Afghanistan with his followers. However, with a number of businesses and followers still in Sudan, he retained an influence there, and many Sudanese companies are thought to act as a front for some of his activities.

Pushing forward in his vision of a fundamentalist Islamic world, in February 1998 bin Laden formed an umbrella organization known as "The Islamic World Front for the Struggle Against the Jews and the Crusaders" (Al-Jabhah al-Islamiyyah al-'Alamiyyah li-Qital al-Yahud wal-Salibiyyin). Its members were to include the Egyptian terrorist groups al-Gama'a al-Islamiyya and al-Jihad, as well as some extreme fundamentalist groups in Pakistan. Its purpose was to take the war to Israel, the United States, and its allies. Bin Laden justified its existence by arguing that these countries were responsible for the oppression of Muslims everywhere. Therefore, to protect the vulnerable, as well as the faith, these countries needed to be taken on and destroyed. Three "fatwahs" were declared, calling on all faithful Muslims to take up the holy war, or Jihad, against the "enemies of Islam."

During the 1990s, bin Laden's name was connected with several terrorist incidents (see below), even though there has never been any direct evidence to link him with the bombings. Other planned incidents, such as an attempted assassination on Pope John Paul II during his visit to Manila and the midair bombing of several U.S. airliners were never carried out.

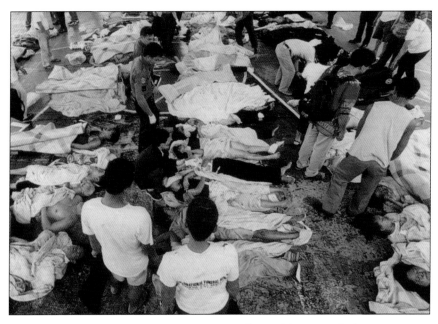

Terrorist incidents produce a lot of dead bodies, mainly among the innocent.

Osama bin Laden Terrorist Attacks

December 1992 – A bomb exploded at a hotel in Yemen, which was hosting U.S. soldiers en route to Somalia to join a UN operation. Several tourists were injured.

October 3–4, 1993 – It is thought that supporters of al-Qaeda took a part in the shooting down of U.S. helicopters, as well as the death of U.S. servicemen in Somalia.

February 26, 1993 – A large bomb exploded at the World Trade Center in New York City, killing six people and injuring over a thousand.

June 26, 1995 – An assassination attempt on Egyptian President Mubarak in Ethiopia was foiled.

November 13, 1995 – An explosion occurred next to a residential compound in Riyadh. Six people were killed, including four Americans, and sixty were injured.

June 25, 1996 – A large bomb contained in a tanker lorry exploded next to the Khobar Towers military housing complex in Dhahran, Saudi Arabia. There were nineteen deaths and 372 wounded, most of whom were U.S. servicemen and women. Although Hezbollah has been blamed for this attack, it is suspected that Osama bin Laden was also involved.

However, bin Laden's involvement in the bombings of two American embassies in Nairobi, Kenya and Dar-es-Salaam, Tanzania on August 7, 1998, was incontrovertibly proved. The bombs that exploded that day claimed the lives of 301 people and injured 5,000 others, most of whom were natives of the countries involved. The bomber, Mohammed Sadiq Odeh, was arrested in Pakistan and yielded a good deal of useful information to the security services regarding the setup of bin Laden's operation. Using the intelligence, the U.S. military managed to intercept bin Laden's communications network and gain the proof they needed of his involvement. With new knowledge as to the whereabouts of bin Laden's training camps and other facilities in Sudan, they used bomber aircrafts to strike at them. However, one of the targets—a baby milk factory—proved to be controversial. The United States claimed it was targeted because it was suspected of being used for the manufacture of chemical weapons, whereas the Sudanese government protested that it was a civilian facility. The U.S. government has since admitted it made a mistake. Since that time, bin Laden and al-Qaeda have been shown to be responsible for two more atrocities:

October 12, 2000 – The USS *Cole* was refueling at the port of Aden in Yemen when a small craft pulled alongside. The suicide bombers in the boat exploded the bomb, causing the deaths of seventeen U.S. servicemen and women and wounding thirty-nine others. A bomb also exploded on this day at the British Embassy in Yemen.

Sept 11, 2001 – Four aircrafts were hijacked while flying out of U.S. airports. Two were flown into the twin towers of the World Trade Center, causing nearly 3,000 deaths. One was flown into the side of the Pentagon, causing 189 deaths, and the other was brought down by the passengers without causing any loss of life on the ground. All forty-four people abroad the aircraft, however, were killed.

USS Cole, an example of what terrorist groups can achieve even against the might of the US Navy.

The attack on September 11 prompted a declaration by the United States that it would rid the world of terrorism and al-Qaeda. Backed by most countries worldwide, it launched a successful attack on Afghanistan, intending to destroy all of al-Qaeda's bases, overthrow the Taliban, and arrest most of the leaders. While the first two goals were mostly successful, bin Laden and those closest to him, including the Taliban leader Mullah Omar, escaped and went into hiding. It is thought that many of their followers are also in hiding, waiting for the right time to strike at America and its allies again. Considering that al-Qaeda has many supporters all over the world prepared to die for what they believe in and considering that the organization may yet have the knowledge, if not yet the capability to produce weapons of mass destruction, this is still, in most people's opinion, the most dangerous terrorist organization in existence. However, even bin Laden himself was not immortal: On the May 2, 2011, shortly after 1:00 a.m. local time, Osama bin Laden was located and killed inside Pakistan by U.S. Special Forces.

Operation Neptune Spear

This was an American military/CIA operation carried out in the main by SEAL Team 6 CIA operatives supported by the 160th Special Operations Aviation Regiment. On May 2, 2011, shortly after 1:00 a.m. local time, Osama bin Laden was killed at a villa in Pakistan. The operation, which took off from neighboring Afghanistan, was ordered by President Obama. Bin Laden's body was first taken to Afghanistan for identification before being buried at sea.

Building where bin Laden was found and killed.

Taliban

We have all heard much about the Taliban and the fighting in Afghanistan, but not all Afghans are Taliban supporters. The Taliban, which is made up mainly of members from the Pashtun tribes, ruled much of Afghanistan from September 1996 to October 2001, including the capital of Kabul. As the Islamic Emirate of Afghanistan, it enforced strict interpretation of Sharia law to the point where many leading Muslims became very critical of the Taliban's interpretation of Islamic law, in particular their brutal repression of women. During their period of power, they were supported by Pakistan Intelligence Services as well as al-Qaeda. The Taliban have committed several atrocities against the Afghan people and during their grip on power conducted a scorched earth policy, reducing much of the fertile land to dust. It is worth noting that prior to the rise of the Taliban, Afghanistan had been in the main self-sufficient in agriculture and crops.

In 2000, British Intelligence reported that Pakistani Intelligence were actively supporting the Taliban, and even when they denied any involvement, after

Barry Davies

Terror in Mexico is on the increase with drug gangs fighting for control of the streets. Assassinations are a daily occurrence.

the September 11 bombing in the United States, Pakistan continued to aid the Taliban by air-lifting their training officers from Kunduz to Chitral and Gilgit in Northern Pakistan.

However, the attacks of September 11, 2001, changed everything as America and its allies sent troops to destroy the al-Qaeda training camps and the Taliban. As the war progressed, the Taliban reformed as an insurgency movement to fight the Islamic Republic of Afghanistan, which is currently backed by the United States. That struggle goes on to this day.

Worldwide Terror

While this book predominantly focuses on incidents that happen around Iraq and Afghanistan, we should not forget that there are also many PMCs working in other parts of the world. The drugs problem in Mexico has gone into overdrive with the various cartels battling it out on the streets with no regard for the safety of innocent people. The situation in Africa has also declined, with more violent attacks on the oil barges and personnel. All these violent zones have a need for protection at all levels, from static installations of value to VIPs. Below is a small insight into just a few of the incidents happening in the world as it stands at the time of writing (2012).

The violence in Africa is unparalleled. This poor human is still alive as police watch helplessly.

Africa

Africa is not a good place to work, either as an ex-pat or as a PMC. While there have already been numerous deaths from shootings, bombings, and kidnappings, it seems likely that the situation will only get worse. Three of the most dangerous Islamic militant groups in Africa are reported to be working together and coordinating attacks that could threaten the whole African region. In June 2012, General Carter Ham, head of U.S. Africa Command, said the indications are that Boko Haram in Nigeria, al-Shabaab in Somalia, and al-Qaeda in the Islamic Maghreb were carrying out cross training and sharing weapons and explosives.

On June 26, 2012, a suicide car bomber killed at least twelve people and injured many more when he drove his car packed with explosives into the Somali military base. The al-Shabaab military group claimed responsibility for the attack—the first incident since the Somali and African united forces (AMISON) took control of the "Afgoye corridor."

Next day, Boko Harem militants killed another fifty people and injured an estimated 130 in three separate attacks on churches in northern Nigeria. The attacks began when a suicide bomber entered the Wusasa Zaria church, killing twenty-four and injuring 125.

On July 1, al-Shabaab militants struck again, this time killing sixteen people and wounding at least forty more when they attacked worshipers in two churches. The attacks were coordinated, as both took place during Sunday morning prayers, with the militants firing at close range and throwing grenades.

Asia

On June 8, 2012, a bus carrying government workers to the city of Charsadda, Pakistan, was blown up, killing nineteen people and injuring thirty-five others. It is understood that the bomb had been

Terror in India as the Taj Mahal hotel burns and 166 people die from the attack in 2008.

placed in the bus prior to leaving the depot. The Pakistan Taliban are thought to be responsible.

A powerful remotely controlled car bomb killed at least eleven people and injured twenty more on June 28, when it exploded beside a passenger bus in Quetta, Pakistan. Nine Shiite pilgrims were among the dead as well as two police officers escorting the bus.

Some eighty-five people were killed in Zamalka, Syria, when a car bomb exploded during the burial procession of a man killed earlier by government forces. The device was detonated as the procession passed the mosque. Anti-government forces blamed the government military for the attack.

South America

Four people were killed and eleven injured when gunmen opened fire on a passenger bus after the driver refused to stop at a rebel checkpoint in Altaquer, Colombia. Colombian authorities blamed the FARC for the attack.

In Mexico City on June 25, 2012, three police officers were killed on an airport forecourt when gunmen dressed in police uniforms opened fire on them. The officers were at the airport waiting to take custody of a man linked to drug trafficking. Drug cartel members were thought to be responsible for the attack. Four days later, on June 29, seven people were injured when a hand grenade exploded in a pick-up truck parked outside the town hall of Nueva Laredo, Mexico. The town, which borders the United States, has been the scene of brutal battles between the Zetas cartel and their former allies of the Gulf cartel.

ELN

The ELN (National Liberation Army) is the second largest guerrilla group in Colombia after the FARC. Fabio Vasquez formed the group in 1963 out of students and left-wingers who admired Fidel Castro's Cuban revolution and wanted to achieve something similar in their country. The movement also attracted many Catholic priests who believed in a radical ideology known as Liberation Theology. The most prominent amongst these was a priest named Father Camilo Torres, who became the organization's leader but was killed during his first fight. He was replaced by a Spaniard, Manuel Perez, who continued to lead the ELN until his death in 1998. The current leader is Nicholas Rodriguez, also called "Gabino." He has been with ELN since his teenage years and has earned a reputation as a tough fighter.

The ELN is more idealistic than the FARC, although both want to see an end to high-level corruption and multinational corporate involvement in stripping resources from their country. They also want to see a fairer distribution of wealth among the peasant classes as well as an end to interference from the United States in their country's affairs. Members tend to split their time and resources between military operations and social work, which may explain why the FARC has become a more powerful voice than the ELN within Colombia (as it has a greatest emphasis on military expansion).

It is estimated that the ELN currently has about 4,000 members but that its support base and territories are constantly under threat by both the FARC, who are considered to be their rivals despite having similar goals, and a third right-wing paramilitary group, AUC (United Self-Defense Forces of Colombia). It has been mainly funded in the past from the proceeds of kidnappings, extortion from the oil industry, and protection rackets. They have also recently moved into the lucrative drug trade. It is thought that the ELN's earlier reluctance to have dealings with drug production was because the former leader Perez had moral objections against it. Up until the 1990s, ELN was also supported, armed, and trained by Cuba and still maintains an office there. During recent times, the Cuban government has tried to negotiate a peace deal between the ELN and Colombian government. However, President Andres Pastrana has indicated that he would rather pool all of his resources into trying to make peace with the FARC.

Despite being included on the U.S. State Department's list of foreign terrorist organizations, the ELN poses much less of a threat to the Colombian government than the FARC. Neither group is responsible for more than 15 percent of the civilian casualties caused by terrorist actions each year. Indeed, more deaths are caused by the AUC who, according to human rights organizations, are responsible for more than 75 percent of civilian deaths in Colombia each year. However, the ELN do kidnap more people than the FARC. In 1999, they hijacked a domestic aircraft and forced the pilots to land on a jungle airstrip. All passengers and crew were then abducted. Soon after, 150 worshippers were kidnapped during a church service in Cali.

Avoid strange back streets, especially after dark, if you wish to avoid being kidnapped.

Kidnapping

I decided to add this section on kidnapping because I think it is just about the worst scenario would-be soldiers of fortune can find themselves. If you are going to get killed or wounded in a firefight, then so be it—at least you will be in a position to extract yourself or seek help. Plus, if you are killed, then there will be nothing more for you to worry about.

Conversely, if you are kidnapped, it is highly unlikely you will be released; and despite all hope, you will most certainly die. Most kidnappings appear to be opportune rather than planned. That is to say that journalists or construction workers who are either driving themselves or are in a taxi are suddenly stopped at an enemy roadblock. In some cases, it may be that the local civilians (taxi driver, police, etc.) have aided the enemy by directing or taking passengers into enemy-held areas.

Once taken captive, most kidnapped people tend to be taken and held in rural areas away from cities. Upon arriving at your destination, there will usually be a lot of shouting and questions in Arabic: These are designed to frighten you more than anything else. The major topic of questioning will consistently be religion and the differences between Islam and Christianity. They will aim to catch you out, so be very careful what you say and how you act. This is what you can expect in the initial stages:

- They will examine in detail any documents you may be carrying.
- Providing you are not a soldier or policeman, you may find that some of the enemy treat you with kindness and provide adequate food and water.
- Almost all those people kidnapped and later released said those holding them gave the impression that they didn't know what to do with them next.
- In all cases, the captors accused them of being Israeli, English, or American spies.

Barry Davies

Your behaviour when kidnapped may just save your life and keep your head on your shoulders.

Behavior When Kidnapped

There is much you can do to enhance your chances of survival should you find yourself the captive of an extremist group. The following actions have been tried and tested during many kidnappings around the world, and they apply to most cases when people have been captured.

- Avoid capture at all costs, do whatever it takes to stay free, but do not risk running from armed men in close proximity.
- If you are captured, be aware of what is happening around you, and if the opportunity arises, check the time and remember the day and date.
- Remember that during the initial capture the kidnappers will be very excited and high on adrenaline.
- If taken by vehicle to another location, try to estimate the journey time.
- Do not talk back unless asked a specific question: Become the grey man.
- If you can see the sun, try to determine the direction you're travelling in.
- You will most probably be blindfolded, but in the event that you can see your captors, avoid eye contact.
- Do not antagonize your captors, and do not get into a religious debate between Christianity and Islam.
- Answer any religious questions as honestly as possible without causing likely offence. If you are a Christian, say you believe in God. While this may anger them, they will also understand and respect that. Do not say you are an atheist, however.
- Always let them ask the questions.
- Do your best not to sign anything.
- Be aware of self-induced pressures, such as loneliness, fear, or lack of sleep.

- When questioned, do not show emotion, try to be strong or appear frightened and confused: Think about their questions and answer them slowly. Be polite, and thank them if they offer you food or water.
- When isolated and alone, make yourself as comfortable as possible.
- When you hear someone approaching, remain calm but become mentally alert.

Barry Davies

Hostage rescue is not a game, despite Princess Ann smiling as she is dragged from the Killing House during her rescue preparation training with the SAS.

Rescue

If you should suddenly hear bursts of gunfire, shouting, and/or explosives detonating, work yourself into a safe position, such as a corner or lying flat on the floor. Remain close to the floor, stay away from windows and doors, and if your hands are bound, push them forward so they can be seen.

If the door bursts open and you hear a voice in a language you understand, identify yourself. Then do exactly as instructed. The treatment by your rescuers may be rough, and you may be pushed and shoved while still blindfolded. Remember, they are risking their lives to come into enemy territory and rescue you. Conform to their instructions, and be guided by your rescuers.

4

VIP AND PERSONNEL PROTECTION

What constitutes a bodyguard? The image we are usually presented with is that of a muscle-bound heavy man who pushes people out of the way so that a rock stars or VIPs can have a privileged passage. In the real world, this is very far from true, although I do know a few that fit the description. In the main, they are well-trained and well-disciplined individuals whose sole purpose is to protect the life of their Principal.

VIP protection can consist of an individual bodyguard (IBG), a bodyguard supported by a personal escort squad (PES), and even a residential security team (RST). The size of the VIP protection will depend very much on the threat level to the Principal: For example, a country's President will get maximum protection, while a leading industrialist may only have an IBG. Whatever the size, the principles of protection remain constant. However, another factor that will determine the level of security needed is the scale of the potential threat. In addition, it is important to know who is behind the threat, as this also impacts the level of security. For example, in a war zone, automatic weapons are fairly easy to come by, so the degree of viciousness and quantity of armaments requires enhanced security measures.

Just because the VIP has a large security escort it does not mean he or she is safe from danger. There are many cases in which VIP security protection has failed. In 1978, the Red Brigade kidnapped the Italian Prime Minister, Aldo Moro, who spent fifty-five days in captivity before being murdered. The German industrialist Hanns-Martin Schleyer was kidnapped in September 1977, and some six weeks later was found murdered in the trunk of a car. More recently, the former Afghan President Burhanuddin Rabbani was murdered. All these people had what they thought was good protection.

Barry Davies

Hans-Martin Schleyer. Despite warning and a police escort, he was kidnapped and subsequently killed by the Red Army Faction.

Author's Note: At 5:25 p.m. on Monday, September 5, 1977, a Mercedes—registration KVN 345—turned into Vincenzstasse in Germany. In the back seat sat Hanns-Martin Schleyer. Graded a Security Risk One, his position thus demanded a police bodyguard. This protection followed closely in an unmarked car, inside which were the bodyguards: Reinhold Brandle, Roland Pieler, and Helmut Ulmer.

As they turned the corner, Schleyer's chauffeur, Marcisz, suddenly slammed on the brakes, stopping so quickly that it caused the bodyguards' car to crash into the rear. The reason for the sudden braking was a blue baby stroller positioned in the road and a car coming fast towards them from the opposite direction.

As the other car came to a screeching halt, colliding with Schleyer's vehicle, five masked figures raced forward and sprayed automatic fire at the two cars. The police bodyguard had very little time in which to react to the onslaught of machine-gun fire, although two of them did manage to return a few rounds before they were finally hit. It seemed initially that Schleyer's chauffeur had not been affected, but when he attempted to get out of the car and assist Schleyer, he was then turned upon by the terrorists. Within seconds, four men had collapsed, spattered with blood, and Schleyer was dragged struggling into a white Volkswagen Microbus waiting close by. This operation by the terrorists was highly professional in both its timing and accuracy, as police have estimated that during the ninety seconds of the attack, the terrorists fired approximately 100 rounds of ammunition, most of these hitting Schleyer's driver and his bodyguards. Those inside the second car were hit numerous times in the head and shoulders; however, Schleyer's driver had been carefully shot through the heart from a range of less than four yards. The plan did not allow for Schleyer to be harmed in any way. Such attention to detail can only characterize the operation as being extremely professional. The kidnappers later demanded that they would release Schleyer only in exchange for members of the RAF (Rote Armee Fraktion) from prisons in West

Two of the Baader-Mienhoff, Andreas Baader and Gudrun Ensslin; Baader shot himself in his cell, and Ensslin was found hanging in her cell.

Germany, including Andreas Baader, Gudrun Enssilin, and Jan-Carl Raspe, three of the most notorious terrorists in history.

However, the German Government refused to give in, and after a failed hijacking based on the same demands, the three terrorists of the RAF committed suicide in their prison cells in Stammheim jail, near Stuttgart. The next day Schleyer's family feared the worst when the Bonn government telephoned to say that a French newspaper, the Liberation, had published a letter from the Siegfried Hausner Commando saying that they had killed Hanns-Martin Schleyer. The full text of the message read something like:

"The corrupt existence of Hanns-Martin Schleyer has been ended after forty-three days. Herr Schmidt, who gambled with Schleyer's life, can collect him from a green Audi parked in a street named Rue Charles Reguy in Mulhouse."

The call from Liberation was received in Bonn at around 4:30 p.m. At five o'clock, the police in Mulhouse found the green Audi, registration HG-AN 460.

The car was located, and at eight o'clock, the car trunk was opened to reveal Schleyer's body. The autopsy, which was carried out by the French, revealed that Schleyer had been shot in the head three times at very close range. Grass was found in his mouth, and pine needles still clung to his clothing, indicating that he had possibly been shot in a remote forest before being dumped in the car trunk.

Principal

The Principal is the person you are protecting. Some Principals cooperate with the personal escort squad (PES), whereas others wish to dominate and dictate how they should act and react. It is up to the PES team leader to negotiate with the Principals so that they understand the role each person plays in providing for their protection. Some Principals will ignore any sensible advice, and some may even insist on driving a car themselves. In such case, it is up to the team leader to organize the protection in the best way possible. When you do have a Principal that cooperates, and one that has an understanding of the risk he or she is under, the safer the PES can make them.

For example, the ideal location for a Principal is to sit in the rear seat immediately behind the team leader, who should occupy the front passenger seat. While this is the ideal situation, you must bear in mind that sometimes the Principals may have friends or colleagues with them who wish to be seated in the same car. Again, the team leader must allow for this. In any case, once a Principal is seated, the doors of the car should be locked immediately and the windows fully closed.

There is a basic rule about windows being open during vehicle VIP protection and this is called the "230 rule." It simply means that if the car is traveling at less than 35 mph, the windows should not be opened more than two inches. Additionally, the Principal and any guests should always wear a seatbelt. In my personal opinion, however, the PES should never wear a seatbelt: Years of driving around the bandit areas of Northern Ireland taught me that.

Likewise, almost all new cars are fitted with airbags, and should you have a slight accident, be forced into a sudden stop, or intentionally rammed by another vehicle, these airbags are going to cause you major problems. I would advise that the airbags in both the VIP vehicle and any escort vehicle are deactivated.

Even in a war zone, some VIPs relish the fact that they are in a large convoy protection party, especially senior diplomats. Too many vehicles in the protection group only serve to muddy the waters, and unless moving at a snail's pace, control is easily lost. My advice is that the maximum number of vehicles in any VIP protection party should not exceed three, with the ideal number being two. This will enable the PES to move swiftly while offering good protection.

In a two-vehicle convoy, the Principal should be sitting in the rear passenger side of the lead vehicle. The lead vehicle must always stay aware of the support vehicle and drive in such a manner as to allow the support vehicle to do its job of protection. For example, the lead vehicle should never cross traffic lights knowing that the support vehicle cannot follow. When driving, the lead vehicle should always stay off to one side, allowing the support vehicle to see the road ahead. Although both drivers should be fully conversant with the route they are driving, the lead vehicle should also provide ample warning and indication of turning, slowing down, or even stopping. Both drivers should also work out a set of safety drills and vehicle positioning to cover all eventualities.

Remember, the lead driver has to always think for the support driver; when changing lanes, overtaking, turning right or left, etc. In all cases, they should make sure there is enough room and no obstruction to the support driver. This means that the driver of the lead vehicle must anticipate the amount of space and time needed for both vehicles to commit any single maneuver.

Be aware that when driving in heavy traffic or on a fast highway, you will need to be thick skinned as your actions may sometimes force others out of the way. Take my advice and totally ignore the "finger" and further abuse and simply carry on.

Whether your VIP protection consists of one vehicle or several, the threats you face in a war zone remain the same. The largest danger comes from hitting an IED, but death and injury can also be caused by an ambush by small-arms fire. Having a vehicle breakdown in a no-go zone is also a very dangerous situation as your location will be noted and may be threatened by a hostile force.

If you are unlucky enough to drive over an IED, then most of your troubles will be over; even if you do survive, the chances are that you will be heavily maimed. In such a situation, you will be totally dependent on any security forces loitering in the area and, God speed, a helicopter close by.

VIP Protection

VIP and personal protection covers a wide range of skills, including knowledge of close protection drills, residence and compound security and control, client transfer, and vehicle convoys. However, when we talk about protection, we are in reality talking about violence: Understand the situation, and you will understand what protection is required. Protection is about you: as a bodyguard or member of a PES, you will need to be able to protect yourself before you are able to protect others, such as your Principal.

Many PMCs supply VIP protection for people such as politicians, media, and directors of blue-chip companies. When operating in a war zone, VIP protection differs drastically from those protection drills carried out on the streets of London or New York City. The threat to the Principal and to the protection team in a war zone is at the top end of the danger scale.

VIP protection can be either on foot or vehicle mounted.

VIP protection involves three elements: compound protection, close foot protection, and vehicle protection. We will deal with each of these elements individually. Additionally, there are two main profiles when it comes to VIP protection: high profile and low profile.

The Principles of Personal Security

Take responsibility for your own safety, and make sure that everyone you are operating with does the same. It is up to you to make sure that you have checked your equipment, maintained your level of alertness (no hangovers), and know the correct drills in the event of an incident.

Prepare and Plan and Practice

Know the likely threats that you will encounter, and plan contingency measures in case they happen, including escape routes.

Make sure that your team are also aware of these drills and practice them as much as possible, making sure that everyone will know what to do. Gather as much intelligence as you can from other operators in the locale, and keep well informed of events.

You need to be on your toes and stay alert when on VIP protection duty even when driving in convoy. All the aerials on the two black vehicles will be a range of high-power HF, VHF, UHF, and microwave jammers.

Stay alert at all times. If you are alert to your environment, you may see the signs (combat indicators) that point to something going down, and there will be less chance that you will be taken entirely by surprise.

The Personal Escort Squad

Usually made up of the more experienced members of the security team, the PES provides the main cordon of protection around the Principal. They can either be part of the inner bodyguard protection or will provide the security that will be the first ring of defense against any threat. In that capacity, the PES will always be hyper aware and ready to either engage the threat in order to allow the Principal and bodyguard section to escape or else give body cover and get the Principal away from the danger as quickly as possible.

Speed and good reactions will save the life of your Principal.

The Security Advance Party

The duty of the security advance party (SAP) is to make a recon of any route in advance of the Principal traveling along it. This could be an overseas destination and done a few weeks in advance, or it could be a trip down the road, where the SAP needs to deploy within minutes of being informed. The SAP will be looking for signs of a potential ambush, IEDs, and also, in venues or accommodation, any signs of hostile surveillance. SAP operatives need to be highly disciplined in keeping a low profile, have good communication skills, know how to recognize potential risks, and carry out efficient search drills.

The Residence Security Team

As its name implies, the residence security team (RST) is responsible for the Principal's security when he or she is residing within any building, whether that be at home, a hotel, or at a business lunch. Duties include an evaluation of the property's security and security systems (and knowing how to use those systems), keeping the property secure from any attacks or intruders, and searching for any hidden hostile surveillance devices, such as bugs or cameras. Time can go slow for those on the RST, but you will still need to remain alert at all times.

Author's Note: It is patterns that get VIPs killed. As the records show, a Principal who leaves his or her residence every day at eight o'clock in the morning and returns at five o'clock in the evening is just asking to be assassinated. Most assassinations incidents happen within 500 m (1/3 mi) of the target's residence/office/hotel because place, time, and location are known to the enemy.

A Palestinian by the name of Zohair Akache was in England, apparently studying aviation engineering. What is known is that he lived in semi-squalor by renting a bed-sitter in the Earls Court area. Described as an intelligent and hardworking man, Akache was somewhat of a loner, although he did have a Yugoslavian girlfriend. Like all Palestinians, Akache came under the scrutiny of British Intelligence, more so when in late 1974 he attacked a policeman in Trafalgar Square. He was arrested and identified as a PFLP (Popular Front for the Liberation of Pakistan) activist, but charges were dropped, and he was released with a warning. In March 1976, Akache was arrested again for assaulting a police officer and was charged and found guilty. He was given a "supervised departure," and Special Branch escorted Akache to Heathrow airport where immigration marked his file "No Entry."

Using a false Kuwaiti passport, he entered Britain again on March 23, 1977, and made his way to London. The reason for his return became predominantly clear when at 11:00 a.m. on Sunday, April 10, the former Prime minister of North Yemen, his wife, and one of the country's diplomats were shot and killed outside the Bayswater

Hotel. The ex-premier, who had been visiting the diplomat at the Lancaster Hotel, left to take the embassy car, which was waiting a few yards down Westbourn Street. As the car was about to move off, a man described later by a witness as of Middle Eastern origin moved across the street from where he had been standing on the pavement opposite the hotel, at the junction of Westbourn Street and Sussex Gardens.

The assassin opened the back nearside door and fired several times at the occupants with a silenced automatic pistol. Taken completely by surprise, the passengers could do little to defend themselves, and in less than fifteen seconds, all three were dead. The assassin made his escape, running off in the direction of Hyde Park, and was last seen by the Lancaster Gate underground station. The Photofit picture issued a day later fit perfectly to Zohair Akache. I caught up with Akache some six months later, a meeting that ended in his death.

A. R. Howell

An example of poor VIP protection.

High-Profile Protection

High-profile protection is rarely used in a war zone, mainly because it pinpoints your Principal as a possible target. A typical high-profile bodyguard is one that everyone sees on TV or in magazines, guarding famous celebrities or politicians: often a massive, mean-looking guy who is adept at intimidating glares. This high visibility acts primarily as a deterrent to anyone approaching the Principal, who will tend to draw fans and curious crowds wherever he or she goes (often through his or her own manipulation of the press for publicity purposes). A crowd of fans in itself may not seem like it should be a threat, but it actually poses a serious crush risk to celebrities as people try to get closer to them. And then there is the very real threat of an obsessed fan or stalker who may wish to cause actual bodily harm.

There are times when a visiting head of state or top politician pays a visit to a war zone, and it is in his or her own interest to display a high profile—more as a deterrent to anyone watching the visit on TV. The security detail surrounding the VIP is very obvious: a cordon of highly alert (usually) men, making it very clear that they are packing weapons and will use them if necessary to protect their Principal. Unlike many of the celebrities' bodyguards, these are highly trained and proficient operators, but their task is the same: to provide a visible deterrent while protecting the safety of the client.

Low-Profile Protection

Low-profile protection takes much training and skill, far more than high-profile protection. Here, the operator takes the greatest care to be as unnoticeable as possible and yet still provide the highest levels of protection for the Principal. The PES will normally be taking on most of the low-profile work, blending into the Principal's environment, yet making sure that every member of the team is always in the right place at the right time to eliminate any possible threat.

Low-profile protection is essential in hostile environments when to get noticed means to get targeted, especially if the area has a high likelihood of kidnap or ambush. At the same time, the operator must also be ready to adopt a more visible role if it is thought that a visual deterrent may work better. As always, it will be experience and training that will enable the operator to make the best decision in the circumstances.

To detract attention, the Principal must also be made aware not to draw any attention to himself or herself in the way he or she dresses or acts. If traveling, the route should be recon'd first—if possible—by the security advance party, and then the Principal should be taken to his or her destination in a vehicle that also blends in well with the locale, i.e., not any flashy, expensive car (unless all the locals have those too of course). The driver should maintain alertness at all times for other vehicle threats/ambushes and not drive in a way that will cause unwanted attention.

Barry Davies

Short-range attacks.

Short-Range Attacks

A short-range attack will usually, but not always, come from within a crowd and may involve a suicide bomber, weapon or grenade, assault with a knife, or even a less harmful missile, such as a rock. The immediate reaction should be to alert the other members of the team by drawing their attention to the location of the threat and the type of weapon. All actions should, of course, be instinctive.

The ultimate aim of the bodyguard is to grab hold of the Principal and evacuate him or her to a place of safety—if possible. If you have a PES, it may also be possible to engage the threat, as long as it does not compromise the safety of either the Principal or any innocent bystanders. Engaging the threat also has the advantage of diverting the attacker's attention away from his or her target, making an exit plan easier.

Long-Range Attacks

These may come from a sniper, mortar, or RPG attack and can be more difficult to handle because the threat will be too far away to be tackled directly and neutralized by the PES. Instead, the primary objective should be to get the Principals out of immediate danger; either by knocking them out of the line of fire (to the side or to the ground) and then covering them or grabbing hold of them and getting them into cover.

If you are wearing effective body armor (and you should be), then placing yourself between the aggressor and the target may be an option, but it should not be your primary tactic. If you get shot, then you will be unable to help your Principal, and it is likely that he or she will be next. Keep him or her low and running as fast as possible in order to get out of the killing zone.

In areas where there is an ongoing threat, you might consider additional protection by employing ballistic shields. There are several new folding shields available that look like a briefcase.

Establishing the Ground Rules with Your Principal

It is fundamental to the success of any VIP protection that the BG and Principal start off with a clear understanding of each other's requirements. When the Principal decides he or she needs protection and start recruiting, it is up to the individual BG or the PMC to direct the dialogue on security, which is often different from what the Principal thinks he or she wants. That said, the liaison between the BG and client on all security matters must involve some compromise. To protect someone fully would be too intrusive on the Principal's business or family life, and therefore compromise through discussion should be entered to provide solutions. Such a compromise should involve what is acceptable to the Principal and what is acceptable to the BG or PMC and should include a discussion on the level of protection needed and balanced against the current threat. It is then for the Principal to say what is acceptable and unacceptable so that the BG can make the necessary adjustments. A good working relationship based on trust, honesty, and most of all, confidentiality, will result in a bond between both the Principal and the BG and, over time, allow the BG to educate the Principal into the habit of changing routines. In a war zone, where the threat is normally high, it is unlikely that the Principals will have their family with them; likewise, they will be living in a protected "Green Zone" area. The high risk and dangers outside the protected area will be all too obvious to the Principal, who will gladly follow the BG's advice.

Profiling the Principal

The more you know about your client, the easier it will be to assess the security risks involved in protecting him or her. There are several systems that can be used to provide a framework for profiling, but one of the most popular is known as the "Seven Ps." Using this system as guidance it will be easier for you to know the things you need to be aware of.

1. People:

   Get to know the people who surround your client and whether those relationships are hostile or friendly. This applies to family, friends, workmates, and employers.

2. Personality:

   Basically, if your client makes enemies easily, then there may possibly be a greater diversity of threats. You may find him hard to get along with too, but remember, you don't have to like him to be professional.

3. Prejudices:

   Prejudices against a particular group, political allegiance, tribe, clan, or racial group may make your Principal stand out from the crowd, so be aware of any hatred he may have stirred up in the past or present.

4. Places:

   Your Principal may be in more danger in one locale than another due to factional tensions or previous actions in that place.

5. Personal History:

   Basics such as date of birth, education, previous addresses, employments, aliases, etc., should be obvious, but also make sure you include knowledge of the client's medical history and medications needed in case of emergencies.

6. Political and Religious Persuasions:

   This speaks for itself, especially in countries where both are causing hostilities.

7. <u>Private Life</u>:

Get familiar with how the Principal spends his time away from his main employment, as you will also need to assess the threat level when he or she is relaxing, whether that entails adrenalin sports, bars, golf, or brothels.

Barry Davies

Specialist drivers.

Specialists

Some people are hired for their special skills, one obvious skill being that of a VIP driver. Defensive and evasive driving is not what you see on films such as Jason Statham's driving in *The Transporter,* although driving through some parts of a war zone can sometimes feel like it. Always keep in mind that your basic aim is to take your

Principal from one place to another without incident and deliver him or her safely to his or her destination. This can be achieved by having suitable vehicles, good security, a planned route, and careful driving. Remember that almost all car accidents are due to human error and not the vehicle, with a third of accidents being caused by rear-end shunts. The best way to avoid any accident is to drive in your own space under total control while always looking out for what others on the road are doing.

A good driver is a competent driver: one who can handle a car efficiently and effectively so that the drive is smooth. A good driver knows what is going on outside of the vehicle. He or she will do this by constantly observing forward and checking the rear- and side-view mirrors at least once every three seconds. Having good observational skills—together with the ability to drive smoothly—is the secret to VIP vehicle protection. Additionally, a good driver will have an excellent knowledge of the area he or she is traveling through.

If you don't already possess them, these skills can be learned from many close-protection schools. Here you will learn all about the vehicle VIP protection skills required, such as embussing and debussing (loading and unloading the VIP and PSE) procedures, evasive maneuvers, ramming techniques, and anti-ambush drills.

Become Aware of Violence

If you are aware that violence can take place anywhere and at any time, then your chances of avoiding it are greatly increased. Just because a situation looks calm and normal, always expect and be prepared for the worst. Spotting trouble before it begins is the best way of avoiding it. For example, if you get on the train and find that at halfway through your journey your carriage is invaded by drunken hooligans, simply get off and wait for the next train. Or if you find yourself in a bar that is also frequented by a trouble-making antagonist, just leave.

It is always easy to perceive the threat to be greater than it really is. On the whole, the world is a fairly safe place, with the bulk of the population being mostly law abiding. The same cannot be said of any war zone, where danger lurks around every corner. Being aware of danger will add to your confidence. Being confident is shown through your body movements, your language, and your eye-to-eye contact. Humans are animals, and like all animals, we learn to recognize whether our foe is stronger or weaker than ourselves.

If someone challenges you, stand in an alert position, listen to what your opponent is saying, look him or her straight in the eye, and hold the contact. There are two reasons for doing this:

1. To make an assessment of your opponent.
2. To show him or her what he or she has to deal with.

Author's Note: I have often found that it is better to say very little when confronted by any would-be assailant. Keeping quiet and standing relaxed can have a dramatic effect. At the first sign that your assailant is backing down, break the confrontation and remove yourself. Do not worry about abuse being hurled at your back, but do listen for movement if he or she decides to chase you.

Understanding that you may get hurt, and taking steps to avoid potentially dangerous situations, is called pre-planning and preparation. This is something the British SAS are very good at. Before they go out on any operation where they may encounter an enemy, they study, they equip, and they plan for contingencies. Protecting yourself against physical attack is a natural reaction.

There is no magic formula to defending yourself, no special technique that the SAS keep to themselves—there is only reality. Your defense is all about reality. If you are a middle-aged business-man and are attacked by several thugs, your chances of beating them are slim. The reality is that they are fit and quick, whereas you, on the other hand, have spent too long sitting behind a desk. You may be an accomplished self-defense expert, but the odds are they will beat you due to their numbers. If you cannot extract yourself by running away or getting close to help, then you have to reduce the numbers in order to stand any chance. To do this you must be fit and confident in your actions. The alternative is to lie down and take the punishment.

Any SAS soldier will tell you that the one weapon the Regiment uses time and again is speed and surprise. Your speed will come from good fitness training and practicing various self-defense techniques until you are proficient. Surprise comes from having the confidence in yourself. Any attacker will normally pick on someone weaker or smaller than himself; he or she does this in the knowledge that he or she will win. You can surprise an attacker by facing up to him or her in a confident manner. You can surprise him or her by preparing to fight. Best of all, you can surprise the attacker by getting in the first blow—one that makes him or her think.

Martial arts, no matter what form they take, all depend on two factors: speed and balance. We need to acquire the skill necessary to overcome any antagonist; to this end there is one outstanding prin-ciple:

WITHOUT BODY BALANCE THERE IS NO STRENGTH.
SPEED WILL GIVE YOU FIRST CONTACT.

The "On Guard" position is the first combat move any new SAS recruit is taught during his self-defense lessons. It is not complicated; it only means standing and moving like a boxer.

Barry Davies

"On Guard" position.

Stand facing your opponent, and part your feet until they are about the width of your shoulders. Favor one leg slightly forward, and bend your knees. Keep your elbows tucked in, and raise your hands to protect your face and neck. It is best to practice this move in front of a large mirror: Stand relaxed, and then, with a slight jump, go into the "On Guard" position. Tell your body it is a spring at rest.

First try using your hands: Throw out your favored hand in a blocking motion, and at the same time automatically place the other hand in front of your lower face. Protect your mouth and nose, but do not obscure your vision. Next, imagine that someone is about to punch you in the stomach. Keep your stance, elbows in tight, and twist your shoulders from the waist. You will find that this puts the muscle of your forearm in a protective position, without having to move your feet or upsetting your balance.

The human body is well adapted to taking punishment and will survive even the worst assault; this is one of the reasons we have progressed to the top of the animal chain. We can live with no arms or legs, without eyesight or hearing, although life is obviously a lot better with them. The most vulnerable parts of the body, and those of your attackers, are as follows:

Eyes

We need our eyes; without them, we are fairly helpless. Damage to the eyes will cause temporary or even permanent loss of vision. Vision will allow you to escape any attacker. [4]

Ears

These are a good target to attack, as they offer themselves readily available for you to bite. Sinking your teeth into someone's ear lobe will have the desired effect if you are being attacked. A long sharp fingernail will also produce a large amount of pain. Clapping your open palms over both your attackers ears will produce a rather nasty

---

[4] Note: blinding an attacker should be a last resort, i.e. when you are about to be killed.

numbing effect to the brain and has been known to cause unconsciousness and painful burst eardrums.

Nose

Like the ears, it protrudes and therefore offers a good target to bite or strike with your fist. Feel free to use as much force as is deemed necessary to make your attacker break off his or her attack. Any upward blow will make the attacker lift his or her head and offer his or her throat for punishment. As with the ears, a sharp fingernail pushed up the nostril will be very painful, or just giving it a good smack.

Neck and Throat

The neck and throat can be very vulnerable; the area contains most of the important vessels that keep us alive. Both main blood vessels that supply the brain are located close to the skin surface on either side of the neck. Damage to either blood vessel can cause death. The airway is also easy to damage, and a simple blow will incapacitate your attacker, allowing you time to escape. A single sharp blow to the back of the neck will cause a temporary blackout.

Stomach and Solar Plexus

A heart punch aimed at the point where the ribs start to separate will have a devastating effect on any attacker. Likewise, as most people do not have a muscle-bound stomach, the same blow delivered with force will literally knock the wind out of a person.

Testicles

Although a good kick or blow to the groin will hurt a woman, it will cause triple the pain to a man. It is also possible to grab and twist a man's testicles: While this procedure may repel some woman, it will produce a lot of pain.

* * *

In a situation where no other weapons are available, you must defend yourself with your body's weapons. Select which is appropriate to the situation, and when you decide to strike, move with all the speed and aggression you can muster. Remember to adopt your On Guard position. Think about your actions.

Barry Davies

Deliver a single blow.

Delivering a Single Devastating Blow

During my life, I have been in the unfortunate situation of not being able to walk away from an attack. But I was lucky, because during my many years with the SAS, I was taught one thing that has seen me through all of them: to deliver a single devastating blow. If through practice you have managed to unbalance your assailant, and in doing so, left him or her open to your assault, take advantage of this and follow through with one good blow. Assess the attacker's position, calculate his or her next action, and move before he or she can recover. If your opponent is down on the ground, use your feet to kick him or her in the kneecap or stamp down on his testicles (if your assailant is male).

Your aim should be to incapacitate your assailants and stop them from chasing after you. If they are still standing and you have your back slightly towards them, you may be perfectly placed to deal them a damaging blow to the stomach with your elbow. If they are facing you and their face is not too far away, strike them at the side of his neck with the pinkie edge of your hand. Use a sharp motion, hand held rigidly, and the arm bent at the elbow to form a right angle. By attacking this way, you have a striking weapon some eighteen inches in length: an arm and hand that is swept round like a scythe. The same movement can be aimed at the temple, but this time striking with the backside of a balled fist. Develop a short sharp jab. Using a tight fist, rotate your arm from the shoulder first before extending at the elbow. Aim to deliver the maximum force at some point beyond your actual point of contact. Do not swing your blow; keep it short, sharp, and hard.

The secret is recognizing the precise moment when you should strike that one blow and make it swift, sharp, and accurate—driven home while your assailant is unbalanced. This movement can be practiced without an opponent; a punching bag will prove invaluable. Done properly, the "devastating blow" will get you out of most situations.

Learn to develop a good knockout punch. It may not work on everyone, but it will make any attacker think. As I have stated before, most street fighting will only last for a few seconds, so getting one good blow in may make your attacker break off the contact and look for easier pickings. You do not even need great strength to deliver a good punch: It is a matter of maneuverability, speed, and timing.

Balled Fist

It is normal for a person to fight with a balled fist. Use your first punch to hit a vital target area of your assailant. Aim for the nose, chin, temple, or stomach. If time permits, fill your hand with loose pocket change or arrange a set of keys so that they act like a knuckleduster; this will increase the effect of any blow. Rain several blows in rapid succession, and then try running off.

Open Palm

Slapping the open palms simultaneously against the ears—either from the back or from the front—will cause damage to your assailant. Using a chopping motion against the side and rear of the neck is also very effective. If your assailant is very young or elderly, consider a vigorous slap across the face.

Heel of the Hand

The chin jab is delivered with the heel of the hand, putting the full force of your body weight behind the punch. When attacking from the front, spread the fingers and go for the eyes. If attacking from the rear, strike the back of the neck just below the hairline for a very effective punch. As the head snaps forward, use your fingers to grab the hair and snap it back quickly. You are less likely to injure your hand with the heel of the hand techniques.

Edge of the Hand

Edge of the hand blows are executed by using the pinkie side. Keep your fingers straight with your thumb extended. Your arm should always be bent—never straight—when delivering this blow. Use a chopping action from your elbow, and have your bodyweight behind it. Cut downwards or across, using either hand, moving your hand outwards with your palm always facing down.

Elbow

Your elbow is a great weapon when you are side on or have your back to an assailant. Jabbing the elbow into your assailant's stomach will almost certainly drop him or her to the floor. If you have been knocked to the ground, try elbowing upwards into the testicles (if your assailant is male). Any well-connected blow from your elbow will give you time to break contact and run.

Knee

Although it is one of the body's more powerful weapons, it has limited mobility, which restricts it to the lower part of the body. However, its battering ram effect can cause severe damage when driven into the testicles or aimed at the outer thigh, causing a dead leg.

Foot

A hard kick is as good as any punch and can be used just as readily. Keep your kicks below waist height unless you have had some special training. Remember, the moment you lift your foot from the floor, you become unbalanced. Although there are a few exceptions, a kick with your boot should be done sideways. By doing so, you will be putting more force behind the blow, and you will, if needed, be able to reach farther.

Teeth

Biting into any part of your assailant's body will cause severe pain and discomfort. The ears and nose are the favorite places to go for, but any exposed skin will do.

Using Everyday Items as Weapons

Magazine or Newspaper

Roll any magazine into a baton and carry it with you quite naturally. Hold it by the center to stab with, using either backward or forward

Barry Davies

Using a pen as a weapon.

thrusts. Hold the end of the baton if you intend beat your assailant around the head. A rolled up newspaper is a great defensive weapon for fending off any knife attack.

Pen

Most types of pen have a pointed tip, which means that they will penetrate the skin if used in a stabbing manner. Hold the pen as if it were a knife, and use it against any exposed part of the assailant's body, such as the neck, wrists, or temple. The harder you stab with the pen, the better the results.

Walking Stick

This item offers excellent protection for the elderly, although it is not uncommon for hikers of all ages to carry a walking stick. The best type is one with a heavy ornate top and a metal tipped, strong wooden shaft. Use the walking stick as you would a fencing sword: Use a slashing action and rain blows at the assailant's head or solar plexus. This is very useful against any knife or bottle attack, where you should slash down hard at the wrists. You may also be able to stop the assailants pursuing you if you can strike their kneecaps hard enough. It is a good idea to have a small strap securing the walking stick to your wrist.

Baseball Bat

This has been a favorite weapon of mine for many years. However, having a baseball bat in the house is one thing . . . carrying it in the street is another. In any event, mark your target area with care, as the bat can easily kill your assailant. Should you find yourself confronting assailants, and you just happen to have a baseball bat in your hand, aim for their limbs, not the head. If you are attacked in your home, concentrate on the assailants' arms, especially hitting hard at the elbows. Do not touch their legs, as this will allow them to escape.

Bottle

The weapon of many street fights, its design could have been made for fighting. Do not bother to smash the end of the bottle off, as this usually results in the bottle disintegrating altogether. Use the bottle as you would a club and strike for the head and temples. The body joints, such as the elbow and kneecap, are also particularly good to aim for.

Author's Note: Some self-defense books advocate using a cigarette lighter to ignite the spray from an aerosol can. This will work but is highly dangerous, as there is more than a 50/50 chance that the can will explode in your hand.

Belt Buckle

Any belt with a good metal buckle will provide a good defensive weapon. Wrap the tail end around your hand several times, then use the belt in a whipping action. Concentrate your attack on the exposed areas of skin, i.e., the face, neck, and hands.

Flashlight

It is common sense to carry a flashlight with you while walking out on any dark night. Additionally, several flashlights should be positioned around the home for emergencies (see Home Protection). Although expensive, the more modern Mag-light type torches are extremely good and make an excellent weapon (the SAS have used them for years). In any attack, use the flashlight as you would a hammer.

Remember, when no guns or weapons are used, the aim of any self-defense is to enable you to extract yourself or others while using the minimum amount of force necessary.

5 »

ASSESSMENT, ASSIGNMENTS, AND CONTINGENCIES

The aim of close protection (CP)—or VIP protection as it is sometimes known—is to diminish the risk to an individual, party, or family who are considered to be at risk from an attack or kidnapping. Examples of this range from the protection of a film crew at risk from everyday hostilities at the front end of a war zone to the protection of an overseas minister visiting another country.

Depending on the level of risk, a CP team can vary in size, from one or two members to a fully equipped team who can provide residential/hotel security, security advance parties (SAP), and protection during moves by vehicle or on foot. It is important that the team leader or operator can assess the risk and be able to recommend the required level of security to make it commensurate with the perceived threat. In recent years, this CP role has expanded to include armed protection and escort in higher-risk countries, including Afghanistan and Iraq.

How and where contracts for a PMC arise and, therefore, how the offer of work to a soldier of fortune materializes, is down to what is required in the restructuring after the war. Contracts come in all shapes and sizes and may start with protection for a nongovernmental organization (NGO) overseeing fair elections, the rebuilding of ministry buildings, etc. However, it is of no consequence, as once a war zone has been settled to a degree where rebuilding can begin, a stream of contracts will start to appear.

These contract offers are normally placed on a variety of government and non-government websites; they can also be published in international magazines or even in the local newspaper. Moreover, many of the larger and more knowledgeable PMCs will know what is coming up and will try to get in first, prior to any contract being advertised. How and when they appear will be governed by either the occupying forces (normally the Americans) or the interim government. A simple contract might read something like:

Private Security Detail

Client: Undisclosed due to contractual reasons

Location: Kabul

Project Value: US $900,000 per annum

Total Staff: Three (3) international and eight (8) Afghan

Anyone involved in risk management will provide a high profile international consultancy with armored vehicles and an Afghan PSD team led by a former British project manager.

The client—one of the largest consultancies in Afghanistan—has two high-profile senior managers who travel between the client's own guesthouse, ISAF (International Security Assistance

Force), and various ministries, where they work on a daily basis. These managers also make frequent visits to other high-profile Afghan officials across Kabul.

The project manager, former personal bodyguard to Prince Abdul of Somewhere, head of the Some Commission Delegation to Afghanistan, will lead our VIP Protection Team, providing our Afghan personnel with the very highest levels of training in personal security, VIP protection skills, and advanced driving, more specifically:

- General threat and risk assessments to ensure recent and comprehensive knowledge of the client's particular circumstance and needs.

- Specific threat and risk assessments to locations and venues in preparation for visits, including recons and route clearances.

- Continual security information and advice to client to ensure client's safety and care at all times.

- Diary management for client to anticipate client's needs and to be able to conduct recons and assessments.

- Training of Afghan staff to international standards in all aspects of VIP and personal protection, advanced driving, and security.

In order to win any contract, you must remember that several other companies will be bidding for the same work. However, it is not just a matter of putting in the lowest price; you need to consider how much taking on the contract will cost you. Take into account that you may need to pay other employees, insurance, accommodation and vehicles, plus a lot more, and at the end of the day, still make a good profit. But before you do all this, the first thing you need to do is a risk assessment.

Barry Davies

Risk management means taking into account the unexpected, such as getting your vehicle bogged down in the sand and making you a sitting target.

Risk Management

If you intend to become a soldier of fortune, the one phrase you will hear more than any other is "risk management." Why? Because this is what you will always be doing—managing risks. Risk management is the very reason why PMCs started and why they continue to grow. Terror and violence are rampant in the world today and none more so than in a war zone. PMCs must have a consistent way of assessing risk to help them decide how much security is needed to protect their Principal, facility, or task. If, for example, your PMC is contracted to guard a facility in Afghanistan, then the process begins with a description of the facility, including identification of any undesired events that could affect the facility. Once all the threats are noted, they must be graded in order of the level of threat, from which it is then possible to make a calculation as to the amount of security required to eliminate the threat. While this is a very simplistic view of threat assessment, it should highlight to the novice what is required. The main components of any risk assessment are:

- Physical security
- Vulnerability analysis
- Security effectiveness
- Consequence (what if)
- Likelihood of attack

Irrespective of the task, be it looking after a Principal or a major oil installation, there is always a level of probability of an undesired event taking place, either through chance or by design. Therefore, a simple chart can be drawn up.

Probability Level	Undesired Event
Frequent	Likely to occur frequently
Probable	Will occur several times
Infrequent	Likely to sometimes occur
Remote	Unlikely but could possibly occur
Unlikely	Very unlikely; it will never occur

While we can use past undesirable events as a guide to frequency, it is also possible to estimate from past happenings the severity level of each undesired event. For example, if a VIP has been the victim of several attacks and survived, while several of his bodyguards were killed, then another attack is ranked as frequent with a severity level set at 1.

Severity Grade	Implication
1	Death or severe damage
2	Severe injury, major damage
3	Minor injury, minor damage
4	Less than minor injury, insignificant damage

So it is possible to start to categorize the effects of uncertainty of risk with regards to any process, which makes risk assessment the cornerstone of most PMCs. Most risk assessment is carried out by the PMC's management, who will examine the threat, understand the vulnerability, and thus prepare a risk analysis.

This is one exceptionally good skill you should learn about before you decide to become a soldier of fortune. While the variety of different risk assessments is infinite, within a war zone and under the protection of a PMC, it is normally limited to either producing a risk assessment for a VIP or for protecting a facility. The latter may be an embassy, a new construction site, or simply protecting road convoys. Understanding the problem and providing simple effective solutions is what makes a good risk assessment. Don't forget that when you write your risk assessment, you also need to add how you will assign individual roles in order to nullify the risks.

Dynamic Threat Assessment

In a hostile environment, a basic risk assessment—although vital—will not be enough. Things can and will change very fast, and risk factors appear that were not covered by the initial assessment. That events will happen to change is a certainty, and you will need to be able to re-evaluate your situation in order to keep your Principal and yourself safe.

This process is known as a dynamic threat assessment: dynamic because it has to be carried out quickly and assertively, and it is an on-going process. However, just because it sometimes has to be a spur of the moment/reactive process does not mean that you jump in without thinking first. In the event of such a situation—especially if you are full of adrenaline—discipline yourself to take a deep breath and quickly assess the new set of circumstances. This should only take less than a second, but it will stop you from reacting in a possibly reckless and disproportionate manner.

First of all, take note of the people in your immediate environment. Are there any indications that any of them poses a threat? Are they reacting to another threat a bit further away? And don't forget animals either: Attackers have been known to place IEDs on animals. Next, turn your attention to objects: Is there anything that may conceal an explosive device or hide a gunman? Are any of the people around you carrying anything—such as a knife or a gun—that could be used to hurt your Principal? The last step is to check the environment. Look for obvious places to plant an IED and for any tell-tale signs, such as wires or disturbed earth. Be aware, too, of potential ambush sites, especially if the route hasn't been previously checked. If any of these potential threats manifest into real ones, then it is time to escape, if possible, or take direct action against the threat.

Knowledge of any potential enemy organizations is critical to any risk assessment. Knowing who they are (what types of attacks they carry out and how frequently) are just some of the points you need to take into account. Once that is done, do not forget the basic and mundane procedures, such as fire, accident, injury, and theft: They must all form part of your risk assessment plan—too often people concentrate too much on the prime risk and forget the day-to-day things that can go wrong.

Theft of every kind and degree is rampant in any war zone: Lack of food, the "have nots," wanting what the rich have, civil unrest, lack of policing—all make it easy for the thief. Always check out what is happening in and around your location or residence. Just because you are in a "Green Zone" does not make you immune from theft.

Basic precautions, such as placing strong locks to doors and windows and the incorporation of a good alarm system, will all help to protect your residence. Fitting simple deadlocks is one of the cheapest and best ways of securing your location. Before fitting your locks and bolts, it is advisable to think about who will have access to your home. Some locks can be complicated to use and may inhibit some of the younger or older family members of your Principal.

> **Author's Note:** While serving in Northern Ireland, I was often tasked to break into various premises for security reasons. To aid me in this, I used a very good set of lock-picks purchased in America. Although I became quite good at opening the locks, I found my biggest barrier was an old-fashioned dead bolt.

Keys Control

If the PMC team is housed together—or in the same location as their Principal—make sure someone is designated as the "key controller." Keys are very important; losing them can cause you a great deal of inconvenience. Having a spare key can sometimes solve the problem, but be careful whom you give it to.

- Don't leave spare keys in an obvious place—have a good hiding place.
- Never tag your keys with your address.
- If you lose your keys and have no spare, change the locks.
- Check all windows and doors before you leave the building and last thing at night.

Alarm Systems

Alarming your residence is by far the safest way of reducing any threat; the problem is choosing the right one. Most systems work from a control box which activates and de-activates the alarm to your requirements. This unit is normally fitted close to the entry/exit point of your home. From this control box, a series of sensors and other devices are fitted around the house. If the alarm is triggered, a visible box, which is normally sited on the front of the house and houses a siren, goes off; in addition, a flashing light is emitted at the same time. I suggest you fit a silent alarm system, which is monitored and controlled by the BG or the operations room if working in a team. Silent alarms, registered to your mobile, and a light on the communications desk, will provide you with a bit of reaction time, whereas a loud audio system will warn off the intruder. In addition to magnetic switches that are used on doors and windows, movement sensors that perceive movement within a given distance are widely used for detecting burglars and intruders. You should also fit several panic buttons around the place: by the front door, operations room, etc. If your residence has a walled garden, then perimeter alarm devices should be fitted, as this is your first line of defense. The secret of any alarm system is to fit a reliable unit, one that is simple to use (don't forget your Principal may have children) and one that is flexible and will provide the security team with full control.

Deliveries

Most Principals and VIPs have a constant flow of documents or parcel deliveries to their residence; this is one area where the security team has to be on their toes. It is fairly easy to obtain a delivery uniform of a major company, so no matter what the deliverers look like, always adopt a safe procedure of handling any items destined for the residence.

The letter bomb, although popular in the late 60s and early 70s, has declined due to better detection devices. The letter bomb is normally directed at an individual or organization and is limited in size. Terrorists will go to great lengths to make their letter bomb look as natural as possible, using sophisticated triggering devices to make sure that they get the right target.

- Make a clear list of expected deliveries, who is delivering, and at what time.
- Always be suspicious of unexpected packets or thick envelopes.
- Check any package that is heavier than normal.
- Check any package that smells of almonds or marzipan.
- Check if grease appears to be leaking from the package.
- Do not bend or open any suspect package.
- Leave it undisturbed and vacate the room; lock the door behind you.
- Inform bomb disposal.

House Fires

Even in a war zone, fire in the residence is a real threat and can happen at any time. A fire is self-sustaining, providing it has both fuel and air. Trap the fire in a room or confined space, and the temperature will rise until the flames reach flash-over point. This is where the fire will consume everything in its path. Flash-over normally occurs when a large fire is given a huge blast of air, such as opening an outside door or window. The speed of a fire can be tremendous and a whole building may be engulfed in only minutes. People trapped in a fire normally die from smoke or toxic gas inhalation and are dead before the flames consume their bodies. Treat all fires as your enemy.

Fire can be detected by fitting smoke alarms. They are inexpensive, and providing the batteries are changed as instructed, will last for many years. Most have a test button, and some come with a built-in light indicator. Do not be tempted to take the battery out just because the cook burnt breakfast and the residence is full of smoke.

War zones often lack any firefighting force, so keeping a fire extinguisher in the kitchen, or where combustibles are stored, is always a good idea. Fire extinguishers come filled with a variety of different contents, each extinguisher color-coded and designed to react against a specific burning material. For example, fire extinguishers will be red and suitable for use on materials such as burning furniture and wood, but it will not be suitable for electrical fires. Make sure you choose the correct extinguisher for your home and are familiar with the operating instructions.

Fire Drills

Many fires start through carelessness or forgetfulness. The residence team should have a routine before they leave or before they go to bed at night. Check the building, especially where a fire hazard might exist. Team members from a PMC often live in a common room and, despite what they are told, continue to make the same mistakes.

- Make sure electrical sockets are not overloaded, i.e., the operations room.
- Check that no clothes are left drying over electrical or oil-fired heaters (common error in PMC team accommodation).
- Switch off and unplug electrical appliances before you exit or at night.
- Extinguish any burning candles.
- Do not leave aerosol containers (fly spray) on top of hot surfaces.
- Do not leave a smoking cigarette, even in an ashtray.
- Do not smoke when you have been drinking late at night.
- Do not smoke in bed or when you feel tired.

During any long-term residence, the PMC team must plan a simple fire drill to protect the Principal and other team members. Check over the house and look at all the doors and windows; imagine where a fire might start, and make an escape plan accordingly.

- Indicate where the fire extinguishers are located.
- Practice escape routes.
- Give individuals specific tasks, i.e., to check on each other if in different rooms.
- Make someone responsible for the Principal and his family.
- Have an assembly point in a safe area outside, and then check off everyone on a list.
- Do not, under any circumstances, allow anyone back into a building that is on fire.

Barry Davies

Simple surveillance may be looking in your mirror to see what is happening.

Surveillance Operations

In the world of terrorism, knowledge is everything; without it, there is very little chance of success. Although knowledge is obtained through a number of sources, the primary method is one consisting of surveillance, observation, and accurate reporting. As with any other aspect of anti-terrorist work, there are several working levels of surveillance; you need to know what surveillance is and how it can help you.

* * *

The bombing of the twin towers and the Pentagon took the United States and the rest of the world completely by surprise. In the aftermath, surveillance methods were put into play not only to go after the perpetrators of this evil crime but also to gather intelligence as a safeguard against any more horrific attacks.

What is intelligence? Intelligence is any information about a specific person, organization, or country. By information, I mean anything and everything.

Any collected information is then processed, producing accurate intelligence, which in turn enables those in power to make a judgment on the appropriate action required. Collecting intelligence involves human resources—like the CIA or MI5—while agencies such as NSA, GCHQ, NRO, and JARIC rely on eavesdropping and imagery. In addition to these agencies, embassies normally have an operating system whereby information of a particular country is established. Then there is the information collected from military sources and operations. Finally, there is open information which is obtained largely by the world's media . . . in some cases, a lot quicker than by the government agencies. Apart from the media, there is normally a head of department, a person who follows the direction of the government. For example, the prime minister may pose a question to his head of security, asking if he may have any knowledge of a recent car bomb attack.

The answer to this question will be derived from gathered information, which is then interpreted and analyzed into intelligence. The intelligence is then distributed to those who decide what course of action is required. While this is looking at the larger picture of surveillance, the principles for those involved in PMC surveillance work remain the same.

Terrorism is all about capability and intention. So in looking for information, we must first discover who has both capability and intent. Capability is: "I know how to make a bomb from just about anything in the kitchen, but I have no intention of doing so, because it's dangerous and illegal." (I have the capability but not the intention.) Intention is: "I don't know how to make a bomb, but as soon as I am able, I will kill my neighbor." (I do not have the ability, but I have the intention to commit harm.) Surveillance is about gathering information about people who have both the capability and the intention. A perfect example is Iraq and its development of chemical and biological weapons; they had the capability and the intention, or so the West originally thought. As it turned out from later information, they had the intention but NOT the capability.

Who collects the information will depend largely on the enemy target and where that enemy is located. It is very hard to drop off the radar in this modern world, and almost everyone is traceable. Areas of terrorist operations can usually be defined to the organization. Individuals within the group use credit cards, mobile phones, vehicles, shipping, or they may just walk down the street. No matter where they go, they can be tracked. Even those rebels who live in remote areas can be found by the use of spy planes, drones, or satellite surveillance.

Author's Note: In my time, British surveillance was at its peak during the years of fighting in Northern Ireland, where information was gathered against the IRA and other terrorist organizations. Northern Ireland was never an easy place to work for the army, and in particular the SAS. Militarily, the battle against the IRA could have been won years ago: All the British government had to do was remove the army, and in doing so they would have removed eighty percent of all IRA targets. Instead, Northern Ireland became the haven of the military intelligence officer. Gathering information, tasking COPs (Close Observation Platoons), and pretending to be "in the know," was all one big game. Unfortunately, it was a game where people's lives were at stake. Today, after many years and the loss of so many lives, common sense and peace have prevailed. Northern Ireland is once more a beautiful country, where people can walk the streets without fear.

Surveillance Cells

Once the SAS had been committed to Northern Ireland, the "NI Cell" (SAS office commissioned with finding solutions to the problems in Northern Ireland) was set up. This task was given to Captain Tony Ball, who joined the SAS as a trooper, having initially joined the Parachute Regiment. He was commissioned and given the task of organizing the NI Cell, which was to become the basis of 14 Intelligence and Security Unit. He did much to establish the undercover work carried out by the regiment in Northern Ireland. Tony was an outstanding SAS officer with a brilliant future but sadly died in 1978 in a road accident in Oman, where he was colonel of the Special Forces. He can only be described as "a man I would go to war with."

Initial surveillance methods used by the SAS NI Cell were extremely primitive and mainly reliant on covert cars, cameras, and the "mark 1 eyeball" (i.e., a human being). Slowly, new equipment started to arrive, mainly in the shape of excellent communications and improved photographic equipment. Although video was available at the time, it was still very much in its infancy.

SAS communications improved dramatically and, in pre-SatNav days, a highly sophisticated spot code system allowed the desk operator on base to plot the position of every vehicle moving around the province. This system not only offered control during covert surveillance operations but it also provided the quick reaction force (QRF) with a clear position should any SAS vehicle find itself in trouble. A good example of this happened in Belfast when a car belonging to 14 Intelligence and Security Group (known as the Detachment) found itself being pursued by four IRA gunmen in the early hours of dawn. The undercover operator radioed in, giving his location using the spot code, and help was dispatched. Unfortunately, before help could arrive, the IRA car sped past the undercover agent and screeched to a halt. The operator later told me the story in his own words:

> I just knew that I had been tumbled (exposed) and that the four men in the car behind me where about to stitch (kill) me. I radioed for back up but, a few seconds later, the car overtook me and broke hard, forcing me to stop against the curb. I had an M10 machine pistol under my seat which was ready to go; however, I decided to put my trust in the Browning 9 mm in my waist holster. The door of my car was armored, so I got out, using it as a shield. I was already on my feet as the first IRA member opened the driver side rear door. The guy was having trouble bringing his rifle to bear, so I shot him from a distance of about 3 m (10 ft). The next head to appear was from the front passenger seat; he also had a rifle. I shot him in the head. It then dawned on me that these idiots were using weapons (all had rifles which are difficult to maneuver when exiting a car) not compatible with this type of ambush.

I shot the third member just as he turned. I then casually walked up to the driver and shot him.

Although early in the morning, the gunfire had brought many people onto the street; but when they saw the dead bodies, they soon disappeared. Shortly afterwards, backup arrived. If nothing else, this true story supports what I said earlier about speed and aggression in any given situation.

Barry Davies
Modern technical surveillance equipment now encompasses nano unmanned aerial vehicles capable of spying from anywhere in the world. The SQ-4 RECON is a new breed of surveillance capability.

Technical Surveillance Equipment

Anything from a pair of binoculars to a satellite can be described as technical surveillance equipment. The military have been using infrared and thermal imaging devices on the battlefield since the early 1970s, and like most equipment, they have evolved.

Infrared energy—often referred to as IR—is electromagnetic radiation that travels in a straight line through space, similar to visible light. The term infrared refers to the frequency of the energy, which is "infra," or "below," the red range of the visible spectrum. As the frequency of the electromagnetic energy increases, the length of the waves decreases. Infrared shares many of the properties of visible light, but its different wavelength has several unique characteristics. For instance, materials that are opaque to visible light may be transparent to infrared, and vice versa. Infrared is less subject to scattering and absorption by smoke or dust than visible light and cannot be seen by the human eye.

Low-light image intensifiers electronically increase the amount of light available at night, even that emitted by the stars. Such devices can be used to work at night and have been developed into any number of applications. Military pilots fly and deliver their payload in total darkness using passive night goggles (PNG) and a similar fitment is used by the SAS for carrying out CTRs. One of these lenses on a camera or video allows for covert night filming. Gone are the days when you could hide in the dark: Infrared image intensifiers or thermal vision will find you.

In addition to seeing our enemy, we can now also listen to their conversation. Both landline and mobile phone monitoring have become highly advanced. It is now possible to listen and record most conversations between two people, and you do not even have to be in the same country. Email and faxes can be interrupted and even have the wording changed, and you would never know.

Video cameras and associated accessories have also become more sophisticated. It is now possible to secrete a camera no bigger than a pin head into most items: pens, clocks, ties, or even cuddly toys. Many such cameras have a built-in transmitter capable of sending the pictures to a remote monitor or recorder.

* * *

These days we rely heavily on remote surveillance devices, such as drones and satellites, and while these are improving rapidly, they still do not produce a full picture. It is not easy to carry out normal car and foot surveillance techniques in countries such as Iraq and Afghanistan, especially for Westerners. However, over the years, some very clever approaches have been made to the subject of surveillance, highly intelligent drones being just one of them. While it is unlikely that any member of a PMC will become involved in the larger drone fraternity, there are now several small drones that are being carried and used for surveillance. These are small portable flying machines which carry video cameras and other surveillance devices. Easy to operate and having a range of several kilometers, they can be tasked to check for IEDs, enter buildings, or simply perch on a roof top and transmit video of the surrounding area. One of the most advanced Nano Unmanned Aerial Systems (NUAS) is the British-made SQ-4 RECON. This 200 g (0.5 lb) drone is small enough to sit on the palm of your hand, yet it is packed with electronics and surveillance functionality. It is fitted with a tilt HD digital camera and full 3D sonar so that it can enter a room without touching anything. It even has infrared IR lighting to observe at night as well as IR landing lights so it can be perched on a building in the dark. Its powerful battery provides a flight duration of thirty minutes, but with the motors turned off, it will be able to supply up to six hours of video transmission. Added to all this, the drone is almost silent, and when flying above 5 m (16.5 ft) cannot be heard at all. But the most unique feature of the SQ-4 RECON is its communication system, which supports 802.11b/g at both 2.4 and 5.8 GHz and has a 500 mW (20 dBm) output and -93 dBm sensitivity, giving significant increase in range. The Ethernet Bridge integrated directly into the drone electronics allows for development of different GCS modules in the future. The manufactures claim the following ranges:

- Building penetration   500 m
- Dense urban area   1,000 m
- Undulating open ground   3,000 m
- Line of sight   5,000+ m

The surveillance role has continued to grow throughout both the Iraq and the Afghan wars, and today there is a myriad of highly evolved technical equipment available. In the recent civil war in Libya, the uprising developed several of their own ingenious surveillance devices with which to locate the government troops of Gaddafi. However, despite all this equipment, the best and most used surveillance technique of Special Forces is still the Close Target Recon (CTR).

Instructions and orders in a PMC are not as strict as in the regular Special Forces, but it is essential that some form of orders are written and disseminated to those undertaking the task. An example for orders relating to a typical CTR follows.

Target Recon

The basic scenario will be for two teams to infiltrate a target area where a known insurgent is believed to be hiding out. The operation will take place in two phases.

Phase 1. The CTR will establish the facts.
Phase 2. Enter the house and capture the target.

Phase 1: Target Recon

Start time at the RV1: Team 1 CTR, 2200 hours; Team 2 Snatch, 0100 hours (1:00 a.m.)

- Each group will be given an identical task with two hours to plan their mission.
- For the target attack phase the candidates will require body armor and should carry weapons.

- Once they have been given their orders, they will have time to plan and choose a number of items for the task ahead, e.g., cameras, night viewing aids, etc.

<u>Ideally</u>:
- They will move by vehicle to the outer edge of Washir (RV1 at GR 123456).
- They will de-bus close to the target house and proceed on foot.
- Upon reaching the target, they are to carry out a CTR, gathering as much information about the building as they can, and if possible identify the target and any guards he may have with him.

Note:

Intel believes the building will be protected by a couple of uninterested guards. Certain areas of the house and grounds may be protected by PIR sensor lights.

<u>**The CTR is looking for**</u>:
- A route into and out of the target building
- Good infiltration points into the building
- Places to secrete a visual/audio device in order to collect further information
- Position and number of any guards
- The type and number of weapons they are up against
- Positive identification and location of the "snatch" target
- Enemy routines
- Position of any obstacles
- The possibility of a four-man team being able to snatch the target

Once the information has been obtained, the patrol will head back the way they came to RV1, no later than 0100 hours. Upon arrival back at RV1, they will meet up with Team 2 and prepare and present a verbal debrief on their CTR. Team 1 will wait at RV1 in support of Team 2.

Phase 2: The Snatch

Start time at RV1: no later than 0200 hours (2:00 a.m.)

- Depending on the results of the CTR, Team 2 will prepare to assault the target house with a view to snatching the wanted insurgent known as Ali Mustafaleek.
- They will be dressed in full assault outfit and carrying personal weapons.
- They will have stun grenades and smoke available (confusion aids).
- As much of the operation will be carried out using night vision aids.

- They should arrive on target at the house using a silent approach.
- They are to pacify any guards outside the house complex and make sure their escape avenue is free.
- They will then breach the house at the best location indicated in the CTR.
- The assault team will move in and clear their way through the building room by room.
- Room combat drills will ensure that the whole right wing is clear.
- The assault team should be aware that there may be other armed insurgents.

- Once inside, the team will locate and confirm the presence of Ali Mustafaleek (HOOD and CUFFS) and prepare an immediate withdrawal.
- They will extract him to RV1 and meet up there with Team 1.
- Both teams will make their exfil (extraction) at high speed.

CTR team ready to deploy in West Africa. The smaller the team, the better your chances of getting in and out undetected.

6

FORCE PROTECTION

Force protection covers a multitude of things within the spectrum of PMC operations . . . but what it basically means is escort duty: escorting materials, ambulances, VIPs, and even peacekeeping troops of a regular army. The scope is very wide, but at the end of the day, it simply means selecting the safest route from point A to B, and riding shotgun in the event of an ambush or stand-off attack.

For this role, you will need knowledge of IEDs and countermeasures, such as RF jamming, advanced mine-resistant vehicles, ambush drills, route planning, and a whole lot of other stuff, including the basics (like vehicle search techniques). But remember, by far the largest threat to any road convoy is the roadside bomb.

There are approximately 60 million mines laid around the world, with many of them being in Angola, Cambodia, Afghanistan, and Iraq. An estimated 1,200 victims are killed every month, of which 75 percent are children. Mines, booby-traps, and IEDs are effectively designed to maim more than kill. Mines are indiscriminate: No one is safe, and no one is immune. The passage of time has done nothing

A. R. Howell

PMC providing protection for a Japanese military convoy and its stores as they cross from Kuwait into Iraq.

to reduce the effects of the legacy that mines have strewn around the world. If anything, modern plastics have made the mines an even greater danger, as they cannot be discovered as easily. Mines come in a range of shapes and sizes, and with a variety of differing purposes. There are landmines, antipersonnel mines, anti-tank mines, and a whole range of detonating devices. These can be pressure, pressure release, tilt, pull, and aerial. Finally, there are IEDs of which the mine can be any size and shape and made from anything that will explode.

When a mine is triggered, the detonation is instantaneous with the explosive charge blasting whatever fragmentation the mine is made of into the air. Some mines, when activated, laterally jump to around waist height before detonating with devastating effects. Other mines, such as claymores, are designed to be triggered by humans when the enemy is within range. These are normally used as a defensive weapon to secure a perimeter or to stop enemy troops from overrunning your position. When detonated, these mines spread a pattern of steel ball bearings to anything between 25 and 50 m (80 to 160 ft).

Why does the military manufacture and deploy mines? They are used to protect military bases and key installations, protect and deny access across borders, and deny the enemy the use of roads or tracks. In some cases, mines buried by military personnel are distributed by mechanical means, and there is no accurate method of knowing where they are placed. Discovering a mine before it is activated and trying to disarm it is fraught with danger, as most are fitted with some form of anti-tamper or anti-lifting devices.

This old artillery round makes a perfect IED.

In a war zone, you can expect to find mines or IEDs just about anywhere. If you think you have located one or are in any doubt, always stop, warn everyone you're with, then retrace your steps and leave the way you came. Report the matter to the military or prevailing authorities as a possible sighting. If someone in your party accidentally steps on a mine, everyone else should freeze. Do not rush to the aid of the victim, no matter how much they scream, as there may be other mines in very close proximity.

IED Prevention

Defeating an enemy IED and mine infrastructure requires a whole army of different approaches. It necessitates the need for good intelligence so that the IED factories can be located and raided. It needs a strong technical element of IED detection and deactivation. It also requires excellent training and dissemination of all that is known about IEDs to the warfighter on the ground. The growth of the robotics industry has done much to alleviate the need for humans to venture close to an IED, and the use of robotics on the battlefield has grown mainly in support of countering the IED threat. An estimated $24.6 billion is now being spent on new electronic techniques to defeat these lethal weapons. But in the meantime, in the absence of a robotic ally, the preventive methods include:

- Protecting the war fighter with better clothing and equipment.
- Improvement in vehicle armor.
- Imaging techniques that can identify or indicate the location of an IED.
- Surveillance of key areas.
- Radio jamming on commonly known triggering devices, such as cell phones, walkie-talkies, and car door fobs.
- Chemical detection of explosives in the hands of local suspects.
- Ground-penetrating radar.
- Deployment of rapid UAV strikes.
- Specially trained search dogs.

Cpl. Anthony Ward Jr.

IED prevention using a dog to sniff out explosive.

While there remains no real answer to the detection of IEDs, other solutions, especially in the field of injury prevention, have progressed. Both the U.S. and British armies have supplied "ballistic boxers" to soldiers working in Afghanistan. Officially known as protective underwear, they are made from heavy duty silk. Silk has been used for centuries as a form of protection, as knights would wear silk garments under their chain mail. Many soldiers in the army of Genghis Khan also wore several layers of silk, which was enough to stop an arrow or deflect a sword.

In addition to silk, some soldiers also have extra protection for the genitals—this comes in the form of a codpiece or cup that fits the genital area (but is very uncomfortable). Extra Kevlar plates can also be inserted to help protect the two arteries that run up the inside of both legs.

While the ballistic underpants will not stop a bullet or large shrapnel, they do prevent a lot of the sand and other debris thrown up by the blast from entering the wound area. It may seem a little frivolous talking about this when there is the possibility of losing two lower legs and an arm in the blast, but the idea is to keep the soldier alive and in a state in which he or she can be repaired. Advances in repairing IED amputations are truly outstanding. Soldiers who would have died just a year ago are now being saved. However, while flesh and bone can be replaced by metal and plastic, the groin area is a different subject altogether. Serious damage up and inside the genital cavity is difficult to repair, and death is almost certain.

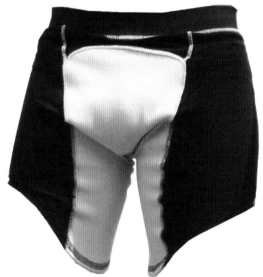

Blast boxers.

BCB International, Ltd.

One company is selling a set of underwear called "Blast Boxers," which are designed to protect against IEDs. Blast Boxers are manufactured in both Europe and America. They are constructed from Kevlar fabric that protects soldiers in areas where standard body armor is absent. Weighing in at just under seven ounces, the Blast Boxers are lightweight and comfortable, as all the stitching is done on the outside. Their main job is to shield the femoral artery, which is located in the thigh and, when severed, produces rapid blood loss. They also help protect against a ruptured colon or loss of genitalia, (which is a major plus in any man's book).

On Discovery of an IED or Possible IED

When it comes to IEDs and bombs, no matter if you are a soldier or a civilian, if you spot something unusual, no matter how trivial it may seem—**CALL IT IN**.

For those soldiers who enter a war zone where IEDs are prevalent, it is almost certain that they will receive good training in the practice of dealing with the situation. However, many others who enter a war zone will not be so prepared. (Personnel working for NGOs and aid agencies may only have an inkling of what to do.) Advice on dealing with IEDs is constantly changing, as more intelligence comes to the forefront. However, I have outlined a simple procedure below, which at least should keep you alive.

- Clear the immediate area.
- Watch which way you move and the cover you take; secondary IEDs may be present.
- Establish a cordon, 300 m (330 yd) for a small device, 1,000 m (1,100 yd) 1 km (2/3 mi) for a vehicle-borne device.
- Request explosive ordnance disposal (EOD).
- Maintain observation of the device until EOD is on the scene.
- Inform your command of the following:

  a. Your radio call sign, organization name, etc.
  b. Time the IED was found.
  c. IED type if known.
  d. Eight-figure grid reference, or close GPS location.
  e. Contact location and route to meeting point.
  f. Any immediate threat to personnel, property, or mission.

Detonated IED

When an IED is initiated, anyone within the immediate vicinity will be caught up in the blast. If the detonation was caused by victim initiation, that person will almost certainly lose at least one limb, more likely two, and sometimes even three or four. The limb that steps on the mine will take the blast, with the other limbs coming in at a close second. The outstretched arm holding the rifle is also at risk, as is the groin area.

Damage caused by an IED includes immediate amputation of extremities, contusions, ruptures, lacerations/punctures/avulsion, fractures, subdural hematoma, concussion (traumatic brain injury), burns, loss of hearing, and flash blindness.

The speed at which the injured soldier can be moved from the point of IED impact to the hospital is vital to his or her life. Where there is more than one victim, effective triage is crucial. Those victims with a good potential for survival should receive immediate medical attention, while those with a poor prognosis should receive minimal care but with maximum pain control. Soldiers with lesser injuries should be checked and placed in order of the degree of injury.

All trauma care should start with ABC (airway, breathing, and circulation). Once the victim is stabilized, as much history as possible

Escort vehicles that are new and bristling with antijamming devices are a dead giveaway to the terrorists as this is just the type of target they go for.

should be obtained. A systematic physical examination should be performed with the patient completely naked and exposed. The findings of the examination will address the priority of the recovery procedure (see Chapter Ten).

Vehicles

Choosing a type of vehicle to travel with in a war zone is a hard call, and the decision should be based on the security situation and the type of work you will be undertaking. If you want to maintain a low profile, then you can buy something local, any old banger, but make sure it is reliable. You might even want to try to armor the vehicle, especially around the driver's seat, or try putting a slight tint on the windows to stop or defuse any profile of you being a Westerner. Do not overdo the latter though, as you will still want to be able to see what is going on outside. Being inconspicuous and making

Barry Davies

Intervention vehicles are mainly used during a building or aircraft siege, but they can also be used as a road block buster if properly fitted.

the vehicle less visible to insurgents is a good thing, but remember that regular military forces might mistake you for an insurgent if you overdo it. If you do come into close proximity to a military convoy, keep your distance—especially when they stop—as all guns will be trained on the surrounding traffic.

If you are going to travel regularly on dangerous roads or through hot areas, then I would personally choose an armored car. However, getting an armored car is not as simple as it sounds, and it costs an absolute fortune. In my opinion, you are better off getting a good car, such as a BMW or Mercedes, and platting it up yourself with armor. A Humvee or ex-military vehicle is even better; don't worry if the finished vehicle looks ugly, just as long as it does the job.

Intervention Vehicles

An Assault Intervention Vehicle (AIV) is designed to deploy members of an assault team rapidly into a target area, which can be anything from an illegal roadblock to buildings. It can be armored and fitted with an assault platform and ladder system. If required, it can deliver up to ten fully armed and equipped personnel to access points at heights of up to 6 m (20 ft) from ground level. In order to provide maximum stability when carrying ten personnel on the platform and side ladders at high speeds, the vehicle's suspension, shock absorbers, and brakes have been up-rated. Differing grades of ballistic protection are also available to protect the vehicle.

The principle idea behind the intervention vehicle is to allow assault mobility, access height, speed, and flexibility during any situation. In short, it is an aggressive delivery system which can put heavily armed men directly into the action. While few PMCs use intervention vehicles, they are growing in popularity for convoy protection, and the system is invaluable when extracting a Principal during an ambush.

Vehicle searching takes place in many war zones; be careful when approaching, and make sure you can manoeuvre if there is an illegal road block.

Systematic Vehicle Searching

No matter what vehicle you drive or where you park it at night, always park it so you can drive away without having to reverse, and check it over each time you get into it.

A full systematic search must be carried out every time a vehicle is left unattended for a period, goes for a service, or is new. This search takes a lot of time and effort, so it should be avoided wherever possible. For instance, do not leave the vehicle in a public place where you cannot see it at any point, as it only takes a few seconds to plant a tracking device or a small bomb. Other possible attacks on the vehicle are the cutting of brake/fuel lines or damaging the tires.

The search should be carried out methodically and should start with the area around the vehicle—in case a bomb has been placed in the immediate vicinity. Next, check over the bodywork, searching for signs that someone has been tampering with the car, such as tool marks or fingerprints. Then proceed to the wheels and arches, and underneath the chassis, looking for anything that has been tampered with (especially brake pipes) or anything that shouldn't be there (look for small boxes or wires where they shouldn't be). Then check the boot area and spare tire before moving onto the interior. This is likely to take some time, as there are so many hiding places for either bugs or bombs. With the interior finished, look under the bonnet and check the engine compartment for anything out of place, as well as levels of brake fluid, hydraulic fluid, water, and oil. Finally, start the engine and check all electrical systems, brakes, gears, and steering. Remember the basics:

- Check your oil and water levels at least once a week.
- Check that the pressure in the car tires is correct, including the spare.
- Check that you have the correct equipment for changing a tire.
- Make sure your windshield wiper is filled during the winter.

Barry Davies
You will need a lot of fire power and luck to get out of a well-planned vehicle ambush.

Ambushes

In addition to the threat from mines and IEDs, force protection also means dealing with both long- and short-range ambushes. The most common form of ambush you will encounter is when the enemy tries to block your route, forcing you to stop. If they succeed, they will then either open fire or try to force you from the vehicle at gunpoint. There are two choices here: First if you are blocked and fired upon, then you should try to extract your vehicle from the area immediately. The success of a small-arms attack will depend very much on the range and accuracy of the attackers, plus the type of weapons they are using. In well-protected vehicles, you will have a good chance of escaping. Secondly, if your vehicle is hit and non-operable, you will need to debus and move on to a safer area on foot. DO NOT stay in the immediate vicinity of the vehicle, because it will be the center of attention for the enemy for several seconds. In such a case, return fire immediately, as much and as aggressively as possible. In those few seconds when you make a run for freedom, gain some running time by giving the enemy at least one more magazine in their general direction.

Once clear of the vehicle, withdraw to a safe location. Under no circumstances should you try to engage the enemy during an ambush, as they hold all the aces. You will NOT know their numbers or what firepower they have. Just get the hell out to some safe area. If this is not possible, hole up in a secure place until help arrives.

If the lead car is hit by an IED, there will be little that you can do for the occupants of the vehicle, and chances are that the enemy will have a back-up plan. No matter how good your route planning, preparation, and precautions for ambush are, when it hits, it will be a shock. Any hostile force using an ambush will choose their ground with care. It will normally be at close range, and they will have you in a killing zone. If you are traveling in an armored vehicle with run-flat tires, you are better off remaining in the vehicle. However, you should watch out for RPGs.

Road Blocks

In many war zones, the hostile force may set up roadblocks. This is a common tactic and is used to kill enemy soldiers and rob supplies (especially weapons). These roadblocks are usually put in a position where you come upon them unexpectedly, and thus those manning the roadblock will be in close proximity to you should they wish to open fire. Such roadblocks are usually sited in contraflow areas, just over the brow of a hill, or round a bend in the road, so that they are undetectable until it is too late. Treat any roadblock with suspicion.

What to do if stopped at a Road Block?

There can be a wide variety of road blocks; some are legal and some illegal, but treat each one with the utmost caution. It is up to the team leader to say what type it is and to take the appropriate actions. If you are in any doubt, remain calm . . . but be prepared to move quickly.

- Bring your vehicle to stop a few meters away from the roadblock so that those in charge have to walk towards you. Observe their movements and physical state. If they appear calm, in uniform, and are not smoking, it is probably a legal roadblock.

- In any event, keep the vehicle in neutral so that it is not stalled but do not apply the handbrake, as you may need to move fast. All should remove their seatbelts.

- If you are carrying weapons such as submachine guns, keep these out of sight and your hands be seen—but be ready to move if required.

- Have any identification papers and letters of movement at hand.

- Always carry a certain amount of U.S. dollars, local currency, or packs of cigarettes in order to bribe your way past.

- Always be polite and courteous, and smile—do not argue.

- If at any time anyone hears anything suspicious within the roadblock, he or she must make the team leader aware.

- Always train for the situation so that when a team leader says go, everybody knows exactly what to do.

Road rage is really nasty and can be a distraction if you are on VIP escort duty. Some countries really hate Westerners and show it on the roads.

Road Rage

You might think that road rage is confined to the Western world, but I can assure you that in a war zone, it takes on a whole new form. Road rage is a new word to many of us, but the problem has been around for some time. The conflict normally starts when one driver does something to annoy another. In recent years, such disputes have escalated into attacks and many people have been killed. In a war zone, you have the added problem of a hostile population and stone throwing as well as military convoys who think your car is local and want to run you off the road because you're a possible threat. Trust me, if you take them on, they will open fire. You can avoid road rage in a war zone by:

- Driving at a safe speed
- Giving clear indications, in plenty of time, of your intentions, i.e., to turn left
- Staying with the general flow of traffic
- Not driving too slow or too fast
- Staying on the correct side of the road
- Avoiding being deliberately obstructive, i.e., letting people into the flow of traffic from a side road, unless it looks like a trap
- Avoiding cutting in front of other vehicles, especially the military

If you are being harassed by another driver, try to distance yourself from him by slowing down—**do not stop**. Never get out of your car—even if you have been involved in an accident—until it is safe to do so. If you are working in any foreign country and you are clearly the "foreigner," here are a few tips:

- Avoid eye contact.
- Do not make obscene gestures, like giving the "V" sign.
- Do not slow down unless you are going too fast in the first place.
- If you have made a genuine mistake, smile and form the words "sorry." This will appease most people.
- Do not stop.
- If stuck in traffic, lock your doors and avoid any confrontation.
- Avoid displaying any weapons until the last minute, or when someone is trying to get into your car, and it is not feasible to drive off due to heavy traffic.
- Prior planning will help to make for a comfortable and safer journey.

Author's Note: Friends of mine were recently in Libya, and while they were driving from the airport to Benghazi by taxi, a car drove by with its passenger(s) shooting out the tires of everyone who got in his way. As I said, war zones have little law and order, and the rules of the gun prevail.

Cpl. Jad Sleiman

US Marines undergoing vehicle driver training for convoy duty.

Vehicle Driver Training

Any form of force-protection work carried out by PMCs must be based on having the best vehicles available and excellent drivers. It is not always easy to procure well-armored vehicles, but it is possible to get very good drivers. Good drivers that have undertaken driver security training will have a good working knowledge of embus/debus drills, route selection and planning, plus all the escort drills involving several vehicles. They should also learn how to fit and operate anti-jamming equipment carried in their vehicle if operating in a war zone prevalent with IEDs.

The majority of vehicle ambushes will either be static or rolling. A static ambush can be initiated by an IED or by some form of blockage—normally another vehicle. The first thing you will know about the IED type ambush is the bomb going off; it may hit you directly or simply be too close for comfort. Your chances of surviving the killing zone of a well-planned and executed ambush will be extremely slim. A well-planned ambush will not only cover the immediate killing zone but will also have stops, both forward and rear. Remember that the aim of any ambush is to trap you in the killing zone, so learn the skills you need to avoid getting trapped.

- Develop anti-ambush driving skills, which means being able to spot any and all unexpected movement directly in front of you or to the sides.

- Spot the blocks. If you see even the nose of another vehicle coming out of a side street, be prepared to accelerate or brake hard. Always be extra vigilant at places where you are naturally forced to slow down or stop, such as level crossings or traffic lights, etc. Beware of terrorists pretending to be police or military at road blocks.

- Practice emergency braking, especially heavy braking while driving at high speed. Aim to keep the vehicle under control on all types of surface.

- Always look for a way out, especially at junctions where an ambush might happen.

- Develop good reversing skills. Learn to brake and reverse in one smooth movement, using only your side mirrors as that offers better body control; look over your shoulder only if the side mirrors are gone.

- Practice handbrake turns until they become second nature.

Author's Note: Driving through unfamiliar streets—especially in rush hour traffic—can be very confusing. If you are not sure of your route and have no one to help you, consider traveling during the early hours of the morning. No matter where or when you park your car, always reverse into the spot. This should allow you to jump in the car and drive off without having to reverse, which is much slower and makes you vulnerable.

A rolling ambush is one in which you are hit while actually travelling, or more likely when you are forced to stop at traffic lights or intersections. The classic case is a fast motorbike coming alongside your vehicle with a passenger on the back.

Rolling Ambush Skills

A rolling ambush is one that happens on the move, usually in the form of another vehicle or motorbike coming alongside and opening fire on you. In many cases the ambush may involve several enemy vehicles all aiming to make your vehicle full of holes. Rolling ambushes can happen very quickly, so practice taking the following avoidance steps:

- Note ANY motorbike within close proximity; steer your vehicle away from it if possible.
- Do not, under any circumstances, stop your car: If anything, speed up, and clear the immediate danger.
- If you do come under fire, use your vehicle to dismount the rider and passenger of any motorbikes.

However, it is not all doom and gloom, especially in a war zone, where the enemy will always try to have an escape route of their own. This is generally in the direction from where they appear, and you could use this to your advantage. You should always try to exploit any weakness in the ambush as quickly as possible before your escape is cut off.

If your convoy is escorting a VIP, then you must make sure that, at the start of any ambush, the VIP and any rear passengers who are not part of the protection squad duck down into the foot well of the vehicle where they should be safe. You can do this physically or by word of mouth, but just make sure that they comply. If for any reason the Principal's vehicle is stopped and unable to continue, it is the bodyguard's responsibility to extract the VIP to the support vehicle or place of safety.

In order to survive any form of ambush, you will need to spot the signs early to avoid getting caught in the ambush zone, and it will improve your chances greatly if you use armored vehicles. It is everyone's responsibility to be alert, especially at danger points when the convoy is forced to slow down: Team members should anticipate that an ambush is about to happen and adopt the correct attitude.

If in the event your vehicles are disabled and you are forced to debus and make a run for it, the chances of escape are extremely thin. While the damaged vehicles may provide initial cover for a few seconds until you get organized, they will remain the focus of concentrated hostile fire. The only protection you will have is your body armor and, if you are lucky, protective shields. For such an event, practice drills that will allow you to run in the best manner possible while protecting the Principal. The bodyguard should always choose the direction of escape, and everyone else should comply. A good bodyguard will decide on the nearest safe cover, shout, "On me," and then take hold of the Principal before leading the way. Depending on your usual drills, you may wish other members of the security team to lay down some suppressive fire for the few seconds it takes to remove the Principal from immediate danger. If the support vehicle has managed to extract itself and is still running, then this should become the focus of your extraction. Again, this drill should be practiced over and over.

Author's Note: In some recent war zones, there have been locations where ambushes take place on a regular basis; these are generally known as "ambush alleys." If you know there is a high likelihood of an ambush taking place along a route, I would suggest that you beef up the convoy's security with a lot more firepower. Firepower—and being alert—are the two things that will get you out of most situations, even a well-planned ambush.

In some ambushes, it may be possible to ram your way through any blockage. This decision must be taken by the driver, who will accelerate around the vehicle, either front or tail, depending on which offers the best escape route. Remember that all airbags inside the vehicle should be disarmed. Ramming is not an easy thing to practice, as even with old vehicles there is a tendency to destroy them very quickly. Remember that ramming should be the last option, for while you may break through the blockage, you may also damage your own vehicle and make it immobile. If you must ram, then make sure the vehicle is in first gear, and accelerate as hard as possible while holding onto the gear stick to make sure it does not jump out of gear. If you anticipate or see a definite ambush ahead of you, and you have the time, you should execute a handbrake turn. Before you do this, confirm the position of the support vehicle behind you. To practice handbrake turns, you need to find yourself a nice piece of open tarmac, make sure the vehicle is moving at around 30 to 35 mph then while swinging the steering wheel hard right, apply the handbrake at the same time. If your vehicle is manual, then you will need to depress the clutch at the same time to avoid stalling. When you are halfway into your turn, release the handbrake and hit the accelerator, straightening the car out in the desired direction. You can do a handbrake turn while going forward or reversing.

Route Planning

Route planning is not just a matter of getting your GPS out and following the easiest or quickest route to where you want to be: In a warzone there are lots of factors to take into consideration. As well as being able to fire a weapon, you will also need to know all about map reading and GPS navigation. While most PMCs tend to stick to the clearest and safest routes, there may be times when you need to travel through a hostile zone. This could be off road or through an area held by rebel forces.

Several factors will determine which route you should take: the enemy, the time of day, and what type of terrain lies between you and your final destination. Weather also plays a role; in good visibility, ground features are prominent, but in poor visibility it is easy to make a mistake. Careful study of a map should provide you with a mental picture of the ground relief, which will in turn warn you of any obstacles, such as a river or marshland. This is something lacking in most GPS navigation systems. Therefore, the ability to read a map and navigate successfully is a skill that is well worth learning.

The correct use of a map and compass is a basic skill that every would-be soldier of fortune can build upon until he or she is fully competent in navigational techniques. Other navigational skills, not dependent on a map and compass, can also be learned and are extremely useful in survival situations. These basic skills can also prove useful if your compass or GPS gets lost.

Barry Davies

It is always worth learning how to use a basic map and compass.

Author's Note: During the endurance march of my selection into the SAS, I lost my compass and was forced to use the miniature one from my survival tin. It was extremely difficult to see it at night, and my fingers got frozen just holding it—but it got me to the finishing point.

Maps vary in size and design, so take care, to choose the right one for the job. The military are usually issued with Ordnance Survey maps that have a scale of 1:50,000. Pilots, on the other hand, are given maps with a larger scale and detailing a wider area. Survival maps, usually issued for operations behind enemy lines during war, are generally printed on cloth instead of paper. Look after your maps, fold them with care, and protect them in a map case or waterproof bag sealed with tape. **Do not write on the map or mark it . . . ever.** If you must have a marked route, draw it on transparent paper and use it on your map as an overlay.

When you use a map, you should orientate it so that the map is pointing towards the route you wish to travel. You can orientate your map by inspection, looking for an obvious and permanent landmark (a river or mountain for example). Find the feature on the map, and then simply align the map to the landmark.

You can also orientate your map using a compass. Pick a north-south grid line on your map and lay the compass, flat, along it. Then, holding the map and compass together, turn both together until the compass needle points north.

* * *

You will often hear the military asking or giving a grid reference or GPS coordinate. When you look at the map, you will see that it is covered with horizontal and vertical light blue lines. These are called grid lines and are 1 km (2/3 mi) apart. The vertical lines are called eastings; these are always given first. The horizontal lines are called northings; these are given after the eastings. Each grid

United States Military

GPS is a fantastic navigational aid, but while it will always give you your location (unless switched off), there are not always maps available.

square is defined by the numbers straddling the left grid line of the easting and the center bottom of the northing. For example, the illustrated grid square reads 9413.

Barry Davies

The compass 1:50000 roamer is placed over grid square 9413 with the corner over trig point 457 the marking show a 6 figure grid reference of 946138, close enough to identify the map object.

A grid reference usually contains six figures. This is worked out in the following way: The grid square is mentally divided up into tenths. For example, halfway up or across a square would be "5." This reference point is then added after the relevant easting or northing figure. To gauge the tenths accurately, use the roamer on the compass or a protractor. The grid reference of trig-point 457 is 946138.

Once you have established where you are and where you wish to go, work out your route. Study the map and the distance. Plot the most logical route to your objective, taking into account the terrain and any obstacles. Divide your route up into legs, finishing each leg close to a prominent feature if possible, e.g., a road bridge, trig point, or even the corner of a forest area. Take a bearing from where you are (call this point A) to the feature at the end of your first leg (call this point B). A bearing gives the direction to a certain point. It can be defined as the number of degrees in an angle, measured clockwise from a fixed northern gridline (easting). The bearing for north is always zero. Most compasses have scales of 360 degrees, or more normally they are shown in mils (thousands of an inch), with 6,400 mils in a circle. Some compasses have both.

GPS (Global Positioning System)

The GPS has really come into its own. This high-tech method of navigation is a worthy addition to accurate navigation. Developed by the United States Department of Defense, the GPS system consists of twenty-four military satellites that orbit the Earth, continually giving out the time and their position. This information is picked up by a hand-held receiver unit on the earth. Receiving and assimilating information from several satellites, the receiver unit is then able to fix a position and altitude at any point on the earth's surface. Most Special Forces and pilots are now equipped with GPS navigational aids.

Receiving units vary, as do their accuracy. A deliberate error, called "selective availability" (SA), was built into the system. This dithers the signals so that only a "coarse acquisition" (CA) can be obtained, therefore reducing accuracy to about 40 m (44 yd). The SA can be overridden for military use by a "P" code, and this gives an accuracy of about 10 m (11 yd). P-code receivers are very costly and are not available for civilian use. All users of GPS systems, however, can experience P-code type accuracy during times of heavy military activity, when the SA is switched off.

The GPS receiver unit searches for, and then locks onto, any satellite signals. The more signals you receive, the greater the accuracy, but a minimum of four will get you into the ballpark. The information received is then collated into a usable form: for example, a grid reference, height above sea level, or a longitude and latitude. Individual requirements for use, either on land or at sea, can be programmed into the unit.

By measuring your position in relation to a number of known objects—i.e., the satellites—the receiver is able to calculate your position. This is called satellite ranging. It is also able to update your position, speed, and track whilst you are on the move, and can pinpoint future waypoints, thereby taking away the need for landmarks.

You can purchase a Garmin hand-held or car navigation system for Afghanistan; the maps are quite good for both on- and off-road travel. There is also a DVD version for control-room tracking. Despite its excellent qualities, the GPS system can be shut down. Additionally, the unit eats batteries, and civilian vehicle-mounted SatNavs tend to have very poor imagery . . . so don't forget your map and compass.

Google Earth

What can one say about Google Earth, apart from that it is a brilliant idea and one that improves navigation down to street level where available? It is accessible on your smart phone, laptop, or iPad and just abounds with fantastic features. True, you need the Internet to get it and maintain the mapping, but there are programs and mobile phone applications out there that allow you to capture screen shots and use them as mapping tiles to cover the area of interest without the need for the Internet.

Spot Codebook

A spot codebook is an alternative to using GPS tracking, and in some cases it is more secure and reliable. PMCs working in an operational

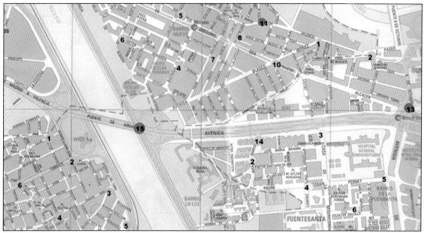

Barry Davies

Example of a spot code map which is easy to make and offers good security if used properly.

area will have several contingency plans in place to help them should they get into trouble. One such basic plan is the spot code system. This allows the desk operator to know the bodyguards' whereabouts at all times. Should they not report in at the given times, a search would be initiated based on the last known location. This is how it works:

A SIMPLE SPOT CODE

A spot code is normally made by allocating a color and number to each major road intersection. As the bodyguard drives from one intersection to another, he or she simply identifies himself or herself and tells the desk operator the code of his or her location. When the area is new to the agent, he or she will carry a spot codebook in his or her vehicle. However, should this codebook fall into enemy hands, it will present a short-term security risk. This will result in all the spot codes being changed, with everyone having to learn the new codes.

The spot code system can also be used in cities when personnel are carrying out surveillance. Additionally, certain spot codes will refer to actions rather than a location; this helps to throw the enemy off balance should they be watching the agent. For example: Sierra Papa four towards black 6. In reality, the agent is simply tell-

Barry Davies

"Cat's eyes" are used to mark a covert pick-up or drop-off point on the side of the road.

ing the desk operator that he is static in a location, say a café or bar, as the word "black" always refers to a static location; the number is irrelevant but in keeping with the normal spot code traffic.

MAKE YOUR OWN SPOT CODE BOOK

Get a map of your local area or download one from Google Earth. Next, purchase a small sheet of colored spot labels from your local stationery shop. Simply stick the spots onto the map at each of the major intersections or streets. Next, number the spots. Depending on the number of people you intend to trust with your spot code book, make a number of color photocopies. It is a great way of staying in touch with the desk operator or meeting other personnel at a specific location.

Barry Davies

"Cat's eyes" template.

Cat's Eyes

During many PMC operations, there may be a requirement to pick up a returning CTR team. One of the best ways of doing this is to allocate a stretch of road onto which you leave a marker. If a CTR team wishes to be picked up at an unspecified location, they simply use a set of "cat's eyes," so called because of the cat's eyes we see on the roads at night.

The team requesting a pick-up will simply inform the desk operator that they require a pick-up between yellow 3 and red 14, if using a spot code. They may have a prearranged time, but more likely they will call for the pick-up only when they are ready. The stretch of road between yellow 3 and red 14 may be several miles long, and this is where the "cat's eyes" come in. The returning team simply plants their "cat's eyes" in the grass at the side of the road, making sure that the head is visible to oncoming traffic. The pick-up vehicle travels along the route between the spot codes until the driver sees the "cat's eyes" reflecting in the grass. At the signal, he deactivates the brake lights and stops with the rear nearside passenger door open.

The team, who by this time is lying hidden nearby, jump to their feet, pick up the "cat's eyes," and get in the car, which drives away. The whole operation takes just seconds. Any other vehicle driving down the road may well see the reflection from the "cat's eyes" but will probably assume that it is a cat, or other animal, in the grass. If the driver of the pick-up vehicle has a car in his or her rear-view mirror, he or she will simply go round the area until it is clean and approach the pick-up site again.

Of course, you will need to adapt this idea a little to fit in with the country of operation: In the desert, for example, you might use two Coca-Cola cans side by side, with the reflective end towards the oncoming vehicle.

Communications

One thing you are going to need when you set off on your route is some form of protection in the event of a breakdown, accident, ambush, or IED encounter (i.e., some communications). Most people rely on mobile networks but in many war zones these have either declined or have been put out of commission altogether or at best the service is intermittent.

You are much better off using a good HF system, which with the right equipment will give you good coverage over a long distance. Codan of Australia makes a range of HF radio that is excellent for long-distance secure communications.[5]

---

[5] www.codanradio.com

7

MARITIME PROTECTION

Piracy has opened up a whole new ballgame for the PMCs, but despite the need to be firm with the pirates, the law on how they are dealt with is still rather vague. This is mainly because of two reasons: confusion over which law applies on the high seas and whether piracy is the same as terrorism. Politically, pirates can claim that piracy is not terrorism as there is no evidence—despite any speculation—of cooperation between pirates and insurgents/terrorists.

Piracy is the act of boarding—or attempting to board—any ship with the apparent intent to commit theft or any other crime and with the apparent intent or capability to use force in furtherance of that act. The term "terrorism" means premeditated, politically motivated violence perpetrated against non-combatant targets by subnational groups or clandestine agents. What rights, if any, pirates should have, where they should be sent for trial, and where they should be incarcerated is still very much anyone's guess.

When you consider that 80 percent of the world's trade travels by sea, you would think that someone would make a decision. This

Most people will disagree with piracy, but you have to give pirates credit for their audacity in climbing up the side of a large vessel.

active sea trade passes through several areas of international waters where piracy is running rife. The main danger areas are:

- Off the coast of Somalia and the Gulf of Aden
- Ivory Coast
- Off the coast of North Africa
- The Strait of Gibraltar
- The Strait of Malacca
- The Caribbean

In 2008, there were some 294 pirate attacks on vessels. This went up to 406 in 2009. Of those 406 attacks, 217 took place off the coast of Somalia and resulted in forty-nine vessels being hijacked and 1,052 hostages being taken. Of these, some sixty-eight people were injured and eight were killed. With the average ransom payment in 2009 being between $3 and $4 million the cost of piracy is estimated to be between $13 and $16 billion every year. This will only increase substantially in the future.

Pirates are generally poor, turning to crime as their only means of income. This is particularly true of Somalia, where some pirates as young as fourteen take part. This is their only source of income, and they are highly motivated, with the first pirate onboard a ship getting the most money. To climb up the side of a large containership is no mean feat, and it takes a lot of courage, but in many cases the pirates are intoxicated or on drugs such as Khat (an amphetamine-like stimulant which is chewed like tobacco) or marijuana.

The first attacks were shore based, and the usual modus operandi was to launch two skiffs, head out into the Gulf, and attack merchant ships as they passed through. In response, in October 2008, NATO deployed one of their Maritime Groups into the Gulf of Aden; two months later the European Union Naval Force (EU NAVFOR) also arrived in the area, but the acts of piracy simply increased. The United States-led Combined Maritime Forces (CMF) established a counter piracy force soon after (in January 2009), but once again, it failed to stop the increase in piracy. This was mostly because the pirates were now using "motherships" in order to extend the range of attacks and select vessels much further from land. These motherships were basically some of the larger ships that had been previously hijacked and were now used as floating bases for the pirates. Today the Somali Pirates have become even more sophisticated and are using quasi-military attack scenarios; approaching from opposite directions while coordinating with GPS/ satellite phones. Their main focus is on targeting "soft," slow, and unprepared vessels. Many of the pirates are heavily armed with Russian-made RPGs, with the faithful AK-47s being their weapon of choice.

Naval patrols have proved to be a deterrent as, under International Law, they have the power to board vessels where piracy is suspected and detain those caught in the act. But these successes are small in comparison to the size of the problem. The seas are vast, and no navy can protect every ship. And when they do come across them, piracy is not generally viewed as an act of war, meaning that only reasonable and necessary force can be used in self-defense. Prosecuting alleged pirates could be seen as a major impending deterrent: Under International Law, any country can prosecute piracy on the high seas. However, in practice few do so, and many suspected pirates are released without trial. The United Kingdom has never brought a suspected Somali pirate back for prosecution. In most cases, the issue of when you can fire at and kill a suspected pirate during an attack is a very grey area. However, military forces have lawfully targeted and killed those who have taken hostages during hostage rescuing missions—even when those pirates have not opened fire.

The thirst for easy money and the vastness of the oceans continue to provide easy picking for pirates, who in their own home countries have become celebrated. In a place where the average annual income is $250 or less, being a pirate can bring vast wealth. Eyl is a town of small mud huts in Somalia, which has become a pirate heaven. The pirates there are now buying luxury SUVs and trying to open up high-class hotels. Women flood in to marry them, looking upon them as modern day Robin Hoods.

Previously, the captains and crew of the ships had no knowledge of how to stop the pirates. However, some countries, including Israel, France, and Spain took the step of placing military personnel

Barry Davies

This image is of Special Forces climbing up the side of a ship when it is stationary; it's not as easy as it looks.

(known as Vessel Protection Detachments, VPDs) on their ships. This avoided some of the problems of using private security companies, and while many ship owners would prefer professional soldiers or sailors, it was simply a matter of logistics: How do you put a detachment large enough onto the ships?

In addition to armed guards, various devices were also deployed, ranging from powerful hoses to sophisticated radars; even putting razor wire around the ship. Nevertheless, it soon became clear that the biggest deterrent was the greater use of private armed security. Since 2008, more than 200 private security companies have been operating in the Somali area. The best of these are usually run by—and employ—former marines or soldiers. British, French, Dutch, or Dane—all are equally good and properly nominated, financed, and trained. However, a lack of agreed international or national regulations raises many questions, for example, what happens if you mistake innocent

Author's Note: Two Italian marines are currently in custody in India following the fatal shooting of two fishermen who were mistaken for pirates. Italy insists that the marines have immunity, as their tanker was flying an Italian flag and was in international waters when the incident happened. But India wants to try the marines for murder under local laws.

fishermen of being pirates and kill them? Do you simply have the right to shoot an innocent man who is unarmed, just because he is climbing up the side of your ship?

While each country must make its own rules, Britain inaugurated the Security in Complex Environments Group (SCEG) in March 2012. Chair of SCEG, Chris Sanderson, announced that standards designed to ensure high levels of quality and professionalism of all private security companies operating in a maritime environment would be in place before the end of 2012. SCEG believes that the high risk level of piracy in the seas off Somalia makes the development of maritime standards and independent accreditation a high priority in order to reduce risks to private security companies and their employees, as well as to ship owners and seamen, and to ship and cargo insurers and brokers. SCEG will also address accreditation for private security companies operating in complex or high-risk land environments, such as Iraq and Afghanistan.

While private armed security guards (PASGs) cannot board vessels and detain suspected pirates, they do seem to be an effective deterrent as no ship with PASGs on board has been hijacked so far. However, there are various legal questions around using PASGs at sea. When can they use force, and to what extent? Who gives the order to use force? How can they transport their weapons legally when in territorial waters or in port? It has even been argued that

The Navy intercepts a vessel suspected of piracy; the main mother ship is used to take the smaller assault craft far out to sea.

having armed men aboard a ship makes it even more vulnerable, as they could escalate the violence should the pirates attack—but there has been no sign of this.

Attacks

Pirate attacks on shipping have been closely examined in order to understand the method of operations used. Because of this, when a pirate attack is suspected, there are specific actions that are recommended to be taken during the pirate approach. It should be noted that the pirates generally do not use weapons until they are within 300–400 yards of a vessel, which gives them valuable time in which to activate her defenses. The ship's first action will be to warn the pirates that it has seen them and is taking evasive measures.

If not already at full speed, the captain should increase it to maximum in order to open the closest point of approach (CPA) and steer a straight course to maintain the speed. At the same time, he should initiate the ship's prepared emergency procedures and sound the alarm. This basically means that all the crew will either go to their allocated positions or to a place of safety. The attack should also be reported to United Kingdom Marine Trade Operations [UKMTO][*] and other agencies around the world that remain on listening watch.

Motherships, most of which have been previously hijacked, are used to carry pirates, stores, fuel, and attack skiffs to enable pirates to operate over a much larger area and are considerably less affected by the weather. Attack skiffs are frequently towed behind the motherships or even carried onboard and camouflaged to reduce chances of detection by naval or military forces. While attacks can take place at any time of the day, most take place during the hours of first light. They have been known to occur on a clear moonlit night, but these attacks are becoming less common.

---

[6] +971 505 523 215 in Dubai.

Commonly, two small high-speed (up to 25 knots) open boats or "skiffs" are used in the attack, often approaching from either quarter or the stern. Skiffs are frequently fitted with two outboard engines or a larger single 60 hp engine. Increasingly, pirates use small firearms and rocket propelled grenades (RPGs) in an effort to intimidate the masters of ships to reduce speed and stop to allow the pirates to board. The use of these weapons is generally focused on the bridge and accommodation area. In what are difficult circumstances, it is very important to maintain full sea speed, increasing it where possible, and using careful maneuvering to resist the attack.

Where practical, the captain should alter the course away from the approaching skiffs or motherships as well as use any sea conditions to expose the approaching skiffs to adverse wind and waves. Anti-piracy countermeasures such as water jets and other appropriate self-defensive measures, should also be activated if present. Somali pirates try to place their skiffs alongside the target ship so that one or more armed pirates can climb onboard. Pirates use long lightweight ladders and ropes or a long hooked pole with a knotted climbing rope to climb up the side of the vessel. It is daring work and very dangerous. If successful, they will generally make for the bridge, where they will force the captain to slow or stop the ship so that the other pirates may board.

When the crew has been well trained and the best marine practices (BMP) have been adhered to, the majority of piracy attacks have been repelled. A copy of BMP 4 can be found at https://homeport. uscg.mil. In general, successful pirate attacks reveal the following common vulnerabilities of shipping:

- Low speed: High speed should be maintained through pirate waters.
- Ship low in the water due to cargo, making it easy to freeboard.
- Inadequate planning and procedures by the ship's captain and crew.

- No pirate alert watch and lack of self-protective measures.
- Ships not heeding piracy warnings.
- Ships slowing down or stopping when fired upon by pirates.

Thanks to the BMP measures and equipment, the fight against piracy has improved dramatically. These include:

- Visible deck patrols and watches.
- Evasive ship action, speeding up.
- Advanced notification of pirate activity.
- Use of high-pressure water cannon to stop pirates climbing up the ship.
- Use of powerful lights to stop a night attack.
- Antiboarding measures, citadel design.
- Long Range Acoustic Device LRAD (150 dB) to provide early warning.
- Use of armed PSC members (the best deterrent of all).

Fired Upon

Pirates normally attack the bridge with direct fire in the hope that the captain will surrender. The use of good protection from sandbags to Kevlar armored jackets and helmets should be readily available for the bridge team. The pirates will also provide covering fire for their colleagues trying to climb onto the ship in order to protect them while they climb and prevent any member of the ship's crew from trying to push them off. In either case, good protection is needed; not just against small arms rounds but also against RPGs. On many ships, the crew will also have a safe location to muster, or even a purpose-built citadel.

A citadel is basically a strong room equipped with good communications, food, and water and a protective shelter with sealed doors so that the pirates cannot enter. If it looks likely that the pirates will gain armed access to the ship, the captain will give the order for ALL his crew to enter the citadel. Prior to doing this, he will also disable the ship's controls so the pirates cannot steer the ship to a port. Once inside the citadel, he will radio for help on the distress frequency.

If armed security is on board, and once the ship has come under fire, they will take the appropriate action depending on their remit and instructions. While the orders to open fire differ from country to country, the basic rule is if they are firing at you, you fire back. Given

Razor wire on the side of the vessel is a great way of protecting the ship from being boarded by pirates.

that any good PSC members will have prepared protective emplacements, and will be well armed and highly skilled, it is obvious why the pirates have not yet successfully attacked a vessel carrying armed PSC members.

Razor Wire

Razor wire is extremely nasty and, if well placed and securely attached around the ship, can create an effective obstacle. The best type is that which comes in rolls and concertinas out in a spiral; a two-layer deep defense on the side of a ship would make it very difficult for any pirate to penetrate: Concertina razor wire coil diameters of approximately 730 mm (2.5 ft) or 980 mm (3.25 ft) are recommended. To climb through this, people would need good protection to their hands, face, and body as the risk of injury is very high. However, razor wire does have one disadvantage in the fact that it can be easily grabbed by the pirates using a hook and line; they then can either try to pull the wire away to create a boarding point or even use it as an anchor to climb up.

It is also possible to use electrified barriers, but these are not recommended for hydrocarbon carrying vessels. If this type of defense is fitted, it should be marked with very clear warning signs for both crew and pirates. The warning should also be in Somali, as this will act as a deterrent to any would-be pirate.

* * *

There have been many improvements in anti-piracy procedures, drills, and equipment. In any combination, these have been proven to work against pirates attempting to board a ship, and where they have made an attempt, it has been repulsed. Deception in the disguise of human-like dummies placed around the ship, where they can be observed by the pirates, can also give the impression of greater numbers of people on watch and act as a deterrent.

While long-range radar is effective, it is still a very good idea to have the human "mark 1 eyeball" acting as a lookout. While doing a two hour on, four hour off watch can be very tedious, it is a necessary

BCB International, Ltd. *Self-inflating vest.*

precaution that needs to be in place to guarantee that any potential pirate attack is spotted long before it gets within attack range. PSC members on this type of watch will require good visual aids, such as high-power telescopic devices capable of both day and night vision. Having talked to one of the PSCs involved in ship protection, it would seem that a life jacket/bulletproof vest is also a good idea. A short wild—but lucky—burst from a trigger-happy eighteen-year-old pirate at over a mile away can still kill you.

Self-Inflating Ballistic Protection Vest

The thing about being at sea is that you might end up in the water. Don't worry, as there is a self-inflating bulletproof vest made just for the occasion. This is a soft armor (200 g/m² Aramid) body vest with integral self-inflating collar plus front and rear ceramic plate pockets. It has a minimum buoyancy of 175 N and the special treatment given to its 500 g/m² (14 oz/yd²) Cordura means that the jacket is waterproof and offers enhanced ballistic, knife, and fragmentation protection. The vest, without plates, weighs 3.5 kg (8.8 lb) and has protection to NIJ level IIIA/UK HG2/KR2. Just what you should be wearing if you are on ship-protection duty.

In addition to radar and barbed wire, there has been a whole host of inventive products to deter the pirates from trying to board or even get near the ship. These include:

Triton Shield

Recent tests were carried out using Horizon Lines LLC—a U.S.-flagged container ship operator serving U.S. domestic routes and

BCB International, Ltd.

Buccaneer.

Asia on the Triton Shield Anti-Piracy System (APS), which uses the most copious resource available to a ship: water. The Triton Shield device hooks into the fire-suppression nodes and sprays a wall of water down the hull of the ship. The salt-water spray makes it harder to climb up a rope or ladder, especially when carrying weapons, and makes it very difficult to see and breathe. The device could also flood small boats in minutes.

With the Triton Shield APS, a ship is also able to determine the difference between a fishing boat and a pirate craft, day or night.

Buccaneer

Another non-lethal counter-measure to pirate attacks, called the Buccaneer, was developed by BCB International. It is designed to be a light-weight air-pressurized launcher to deter and defeat small fast sea craft at long range. Capable of reaching ranges of up to 850 m (0.5 mi), the Buccaneer's specialized projectiles layer entangling net-ted lines across the surface of the water, disabling outboard engines as they wrap around the prop. The Buccaneer can be used from shore locations, such as harbor installations on offshore platforms, or from on board ships.

The Buccaneer is extremely flexible. Mounted on a slew ring, it can be manually directed to face any area of threat through 360 degrees. Alternatively, it can be locked in place and operated remotely from up to 300 m (328 yd) away (from the bridge or other safe area, for example). Alternative projectiles, such as high visibility smoke and splash projectiles can be used for maximum deterrence. Recovery lines, individual lifeboats, lifejackets, and buoyancy aids can also be launched from the Buccaneer. Properties of the Buccaneer system are:

- Gimbal and slew ring: 360-degree arc
- High-visibility smoke and splash projectiles
- Entangling-line projectiles
- Magazine system for multiple successive use: quick to reload

The latest (fourth edition) of the BMP (best management practices for protection against Somalia-based piracy) has been finalized after a process involving key industry, government, and military bodies. The document contains suggested planning and operational practices for ship operators and masters of ships transiting the High-Risk Area (currently defined as an area bounded by Suez and the Strait of Hormuz to the North, 10° S and 78° E). The BMP4 advice on vessel movement and registration is compulsory under Norwegian regulations as of April 2011. The new BMP4 has been published by NATO Shipping Center and others.

Main Purpose

The main purpose of BMP 4 is to offer advice to ships in order to assist them to avoid, deter, or delay piracy attacks in a High-Risk Area. Experience and data collected by naval/military forces show that the application of the recommendations contained within this booklet can, and will, make a significant difference in preventing a ship from becoming a victim of piracy. Implementing the BMP guidance does not constitute a guarantee against becoming hijacked, but it remains a fact that the large majority of hijacked vessels did NOT implement BMP advice.

What's New? Better Structure and More User-Friendly Layout

BMP 4 does not contain many surprises, and most of its recommended measures were included in BMP 3. However, the document has a more user-friendly structure (including highlights on fundamental issues), updates on the piracy threat, and improved graphics. Advice is given on situational awareness, protection, pre/during/post-attack issues, and reporting. Some of the planning sections have been condensed. A brief one-pager (aide memoir) lists key measures to avoid being pirated.

* * *

BMP 4 contains a reference to the (increased) use of armed guards on board vessels but without recommending or endorsing such use (describing it as "a matter for individual ship operators to decide following their own voyage risk assessment and approval of respective Flag States"). Military vessel protection detachments—if available—are to be generally recommended over private armed guards, according to BMP 4. The document also contains a reference to IMO MSC circulars on this issue (IMO interim guidelines as of late May 2011); see www.imo.org.

8

WEAPONRY AND EQUIPMENT

If you made it this far and you have employment as a soldier of fortune, you are now all set to travel. Before the company sends you into your theatre operation, the first thing you are going to need is some healthy personal protection, i.e., a weapon and some body armor. Unlike your regular army counter-parts who get issued everything they need, you will need to acquire your own.

I know a lot of PMCs, and very few actually hold a stock of weapons; if they do, it is more than likely an illegal stash. So don't think that it will be as simple as your being given a weapon and then getting on a commercial aircraft—that's never going to happen. At best, you may receive a weapon from the company once you have arrived in theatre—but don't count on it.

During the war in Iraq, PMCs were required to obtain an import license from the State Department before they could ship weapons into the country. This license was often turned down, and many PMCs were forced to find alternative ways to arm their operators.

Luckily for them, there was a vast range of Russian-made weapons to choose from, mainly sold by Iraqi civilians. Another option was to know of a PMC that had finished its contract and was departing the country, as it would normally sell its weapons on to another PMC.

Barry Davies

Equipment used by the standard soldier of fortune differs greatly. This guy has just about the minimum required, but it does not look very professional.

The good news is that, while most body armor and ballistic helmets require an export license, you are permitted to carry a full set without a license for your own personal protection. While you can acquire these in-country, I would advise you to purchase a good well-fitting set of protection before leaving. In addition to protecting yourself, you will need to know how to operate a variety of weapons plus some standard military equipment, such as detection and surveillance systems. If you are not familiar with these, then I suggest you learn by checking out the various items on the Internet. In situations of war and violence, there is one thing every soldier of fortune needs: equipment. Much of it is straightforward regular army issue, but more complicated tasks, such as those confronting IEDs, will require special equipment.

> **Author's Note:** To the uninitiated, there is a vast wealth of military knowledge to be found in military manuals. These can be purchased from military stores or downloaded from the Internet (Skyhorse Publishing has released many of these government manuals to the general public). But beware that many are full of "bullshit" written by military staff that have never been in war. That said, if you are going to work as convoy protection, then you might look for a military training manual such as FM 3-24 on counter insurgency, or if you are going to be on convoy escort, you might learn something from FM 4-01.45 on tactical convoy operations. As a word of warning, while you can learn much from these manuals, there is nothing to match doing a good training course on the various subjects.

Barry Davies

Military exhibitions are a great place even for retired soldiers because they enable you to keep up to date with the very latest equipment available.

Military Exhibitions

One of the best ways to understand modern weapon technology and other protective methods first-hand is to visit military exhibitions. These can be found in most countries, and one of the largest in the world is held at DSAi in London; another good one is AUSA in Washington, D.C. If you need to get up to speed on your ground robots and aerial vehicles, then you will need to visit AUVSI, which holds events worldwide and throughout the year. Another good place to find military exhibitions in your country is to look at www.tradefairdates.com/Military-Defence-Exhibitions-Y244-S1.htm. Two others worth a mention are Eurosatory in France and DSA in Malaysia, the latter being especially good for weapons produced by Russia, China, and Pakistan.

Almost all military exhibitions have restricted entry, which means you must have a reason for being there. If you are a military manufacturer exhibiting your products, in the military, or a military supplier, it will be simple to gain access. If you are none of these, simply make up business card with a defense industry theme, and add your name on it. Make sure it is the same name as your passport or driving license, because the security at all military exhibitions is very strict.

The beauty of attending a good military exhibition is that you will see the very latest state-of-the-art military hardware. In many cases, the manufacturers will produce brochures and basic technical information about their products. There is a complete range of products on view, including small arms, pistols, grenades, mines, and counter-IED devices.

Once you have gained access to the military exhibition, you will be presented with an exhibition catalogue. Take ten minutes to select a number of stands that are of interest to you. The layout of most exhibitions is on a grid system and strategically placed in each of the halls to indicate which stand is where. Many of the individual staff will also be more than helpful and will happily supply you with information on their product. Just flash one of your phony business cards, and you might even get a demonstration.

Many of the exhibitions are country specific: that is to say, Jordan will show off the weapons made in that country, whereas exhibitions from the USA will display the latest models of say Whilly X military sunglasses. If, as a soldier of fortune, you know you will be operating in a specific country, it is worthwhile to visit their booth. For example, before traveling to Afghanistan, I would check out the Russian, Chinese, and U.S. trade stands. This will allow you to see first-hand what small arms that country is using as well as supplying to other armies. Many of the sales personnel are ex-military, and most will have served in a war zone; so if you have absolutely no experience, stop and have a chat. They will provide you with good intelligence on what is happening within that country.

Protective Clothing and Armor

When it comes to clothing, you will need to dress for the task, i.e., if you are on VIP protection duty you are not going to need knee and elbow pads unless you plan on taking your Principal into the heartland of hell. If you are not sure how to dress for the occasion, my advice is to be a little conservative. After all, you don't want to turn up and be ridiculed. With luck, the PMC employing you should provide some indication and possibly a list of kit and equipment. It is always worth asking if they have any photographs of the guys in action so that you can see how they are dressed and equipped.

There is a massive industry in paramilitary equipment, and you can buy just about anything you'd like. If you wish, you can easily look like the traditional mercenary, with unkempt hair and full beard while wearing full Special Forces attire. Some PMC operators love this overpriced warmonger look, but it is probably best to avoid the full SEAL Team 6 look if you want to retain some credibility.

A. R. Howell

Standard soldier of fortune dress. These guys are on escort duty.

Go to any of the better websites and carefully select what clothing you need. One word of warning: Try to avoid traveling to your final destination in full war gear, as it will only cause you a lot of aggravation if your aircraft has a stopover in London, Dubai, Delhi, or Kuwait. Immigration does not take kindly to letting "Rambo" into their country.

The best places to purchase your clothing and personal equipment are either a military surplus store or an online shop. You are going to need the following: clothing, boots, backpack, sleeping bag, personal

Author's Note: Most people heading for Afghanistan, such as mercenaries, journalists, and businessmen, will ask about body armor. Don't worry too much about this though, as the biggest danger you will face is the local drivers. You are in more danger of being killed on the road then being shot. Trust me, most local drivers in a war zone are morons.

body armor, helmet, plus lots of bits and pieces. For Internet shopping, try the links below. Some are from the United Kingdom and some from the United States, but there are thousands of others.

www.sofgear.com

www.sofmilitary.co.uk/security

www.americanspecialops.com/equipment/

www.britkitdirect.co.uk/index.html

www.bcbin.com/

Equally as important to the above kit list is what you will be wearing in relation to the prevailing climate. Many people think of Afghanistan as a hot, barren country . . . and in some places it is. However, other parts support lush high mountains covered with snow. The ideal clothing for here should protect and keep your body at a constant temperature. Blood flow helps to distribute heat around the body, so be aware of any tight or restrictive clothing that may hinder this blood flow. In the case of gloves and socks, if you are wearing more than one layer, make sure that the outer layer is comfortably large enough to fit over the inner. If you find yourself overheating, first of all loosen the clothing at the neck, wrists, and waist. If this isn't enough, start taking off your outer layers of clothing, one layer at a time. As soon as you stop exercising or working, you should put these clothes on again, or else you will become chilled. The most important points to remember are:

- Keep clothes clean.
- Avoid overheating and sweating.
- Keep clothing dry.
- Repair defects immediately.

Hydration Systems

A supply of water is vital—especially in the Middle East—and you must carry and drink it constantly. In a combat situation, your mouth will feel dry most of the time, so if you get the chance and your supply of water is plentiful, take advantage of it. It is not just water that's good for you: Experienced soldiers often brew up whenever they get the opportunity. This is because they know that a cup of coffee or tea helps keep you cool. Under normal circumstances, the body is constantly losing water from its natural functions of breathing, urination, excretion, and sweating. In hot conditions, sweating will increase, and so will your body's need for water. Therefore, in any conditions where water may be in short supply, the first priority is to carry as much water with you as you need and to conserve the water already in your body.

BCB International, Ltd.

Hydration system with water cooler attached.

Boots

A healthy pair of feet require good footwear, one of the major requirements for those intent on working in a war zone. This is especially true when walking over rough terrain. It is important that you look after both your feet and your boots. Failure to do so could easily mean that you become incapacitated, no matter how fit the rest of your body is. Always wear clean, dry socks, and be aware of the condition of the skin around your feet. Remember, if you were ever to

Cpl. Timothy Lenzo, 1st Marine Division

Properly-fitted, well broken-in boots are an essential item in a war zone.

find yourself running for your life, your feet may be your only form of transport.

Likewise, if you intend to operate in the Konar Valley of northeastern Afghanistan, where you will be surrounded by high snow-capped mountains, you are going to need a suitable pair of boots. There are so many good boots to choose from today that it is difficult to advise on which are the best. What I would say is that you buy your boots some weeks prior to going into a war zone and break them in. Breaking them in includes wearing them for at least eight hours a day and doing some long walks over rough terrain. Wearing them around the house or to your local bar does not constitute "breaking" them in. It is always worth asking other members of the PMC that are already employed in theatre what boots or footwear they recommend. There are a few points to consider when trying out your boots.

- Pay attention to the underfoot when walking over rocky ground. Can you feel the rocks under the soles of your feet?

- Does the side support feel as if it is providing support?
- Do they chafe at any point? If so, can you fix it?
- How do they feel when wearing thicker socks?
- Did you get blisters?
- Can you get them on and off quickly?

Choosing your socks is just as important as choosing your boots. Once again I would advocate that you select a variety of socks from your local trekking or hiking stores, as they normally stock a good range. You can never have enough socks, and I would suggest you purchase at least six pairs to take with you. Remember to wash them at every opportunity, as dirty socks will ruin your feet a quickly as badly fitting boots.

Cpl. William Skelton

Body armor with ceramic inserts.

Body Armor

A bulletproof vest can help save your life, but be aware that it only covers and protects the chest area. Also, not all vests are equal: The level of protection that the vest offers will determine what type of bullet or shrapnel it will stop. Vests that are reasonably light and comfortable will stop a bullet from most pistols and protect against small shrapnel or a knife attack. Inserting armored plates will help raise the level of protection a vest offers, but be aware that no body armor will stop a heavy .50 caliber round. Added to this if you are hit, even while protected, the force of the round hitting your chest is horrendous and can still cause serious damage.

When choosing your protective vest, make sure it is a comfortable fit, because you may have to wear it for long periods of time. Don't get carried away with all the new developments—stick to using equipment that has been tried and tested in the field. An example of this is the hype that is evident concerning Pinnacle's "Dragon Skin" versus current standard issue body armor.

If you have never worn a protective vest before, you will find it stiff and it will chafe around the neck and arms. Get used to it, and always wear your vest as high up to the throat as possible. Any hard, chafing spots can be dealt with by working on the fabric to smooth out the bits responsible. Try sitting down and standing up, driving a vehicle, and getting in and out of a vehicle while wearing your vest: Get used to the way it feels. Use the vest without its protective plates for a while, and then insert one plate at a time, starting with the front plate. Once you are happy, try inserting the rear plate. You will find a huge difference in weight with the plates in, but do not be tempted to remove them, unless you are sure it is fairly safe to do so.

Not all protective vests are heavy. Some are designed to be worn under your shirt, but these offer only light protection. Basically, the rule is to decide what level of protection you need to safely do your work before you start the job.

Never machine wash your vest; lightly brush any hard dirt off, and then use a damp cloth and a small amount of soap to remove any marks. It will generally come with cleaning instructions, so make sure that you follow these to maintain its effectiveness. Most vests are made of Kevlar, which is degradable by any form of bleach, and the ceramic insert plates are susceptible to ultraviolet light (sunlight). You should handle the ceramic plates with care and not drop them onto hard surfaces, where they can crack.

At night, place your vest in a position where you can pick it up in the dark if need be, and learn to get it on quickly. You will find that most modern vests are built in two halves: the back and the front. These are connected by Velcro closures on the shoulders and at the underside of the arms. When you take your vest off, undo the left or right hand underarm only and slip it over your head. Leave the open side exposed, so you can easily slip it back over your body in an emergency. A typical good vest is described below.

Exo Flex Body Armor

This unique body armor offers superior comfort and freedom of movement while maintaining high-quality ballistic protection. It is a fully articulated design which features a pivoted shoulder harness that allows free, independent movement of the shoulders and arms. It is cut short to allow free movement around the lumbar region, offering an all-around flexible protection solution. The independent movement of the pivoted harness gives better efficiency and speed in limited spaces; consistent weight distribution through both shoulders provides less muscle fatigue and improved ergonomics, as there are adjustment points at the midriff, upper torso, and shoulders. This accommodates more body shapes and sizes, while internal pads space the armor away from the body for ventilation; these can be removed for a low profile. The Exo Flex provides protection to:

- Level IIIA soft armor.
- Level IV armor plates, up to 33 cm (10.5 in) x 26 cm (13.25 in), front and back.

Ceramic Plates

These are heavy but necessary in a war zone. The Level IIIA ceramic plate provides protection against high-velocity ballistic threats up to and including 7.62 mm x 51 U.S. M61 ball armor piercing and 7.62 mm x 54 Soviet heavy ball (Steel Core) ammunition when worn with body armor. Weighing 2.5 kg and measuring 250 mm x 300 mm, the plate can be worn in the front and rear of the body armor.

Cpl. Ryan Rholes

Assault vest.

Tactical Assault Vest

Not to be confused with bulletproof protection, a tactical assault vest provides additional capacity for carrying equipment. Manufactured in a variety of materials, the vest incorporates pockets and pouches for further items of personal equipment, such as spare magazines, personal medical pack, GPS, and flashlights. The vest is of a waist-coat-style configuration, fastened at the front, and can be adjusted for size via elasticized lacing on each side. Do not use the Tactical Assault Vest as a substitute for your body armor, as it offers NO protection.

Gloves

Assault gloves are designed to provide protection for the wearer's hands while still permitting full dexterity of the fingers. Manufactured in black Kevlar/Cordura flame-resistant and waterproof fabric, many tactical gloves are fitted with a special trigger finger. Even in a hot, arid country, I would advise you to wear a good pair of tactical gloves.

Australian Military Services

Elbow and knee pads are well worthwhile and do offer good protection.

Elbow and Knee Pads

Elbow and knee pads are simply a form of protection for when you need to fall to the ground. They offer a degree of comfort and cushioning from sharp stones, or when climbing through rubble.

The pad is designed to flex with the knee or elbow and is held in place with elasticized straps that can be adjusted for comfort. An internal foam ledge stops the pad from falling down the leg or arm. Most are lightweight and one-size-fits-all.

A. R. Howell

Sleeping accommodation can be rough. This is a real PMC base.

Sleeping Bags and Sleeping Mat

The one thing you will definitely need is a good sleeping bag and sleeping mat. Modern technology has produced very small sleeping bags that still offer a large amount of warmth and comfort. Place one of these on a foam sleeping mat, and you are guaranteed a good night's sleep anywhere. The sleeping mat provides insulation from the cold ground and a degree of comfort if you are using it outdoors. Even if your PMC provides accommodation, the chances are you will need to take your own bedding, so choose your sleeping bag with great care as you will be using it least eight hours a day. While it may be hot in the summer, the winter months in Afghanistan can be freezing.

If your PMC has provided accommodation, you might also want to take a couple of lightweight flannel sheets. I would also recommend that you have a rip-stop nylon "zoot suit" for when you are off duty, which you can also use to sleep in.

Backpack

You are going to need a good quality backpack, both for traveling and, in many cases, for storage of your equipment. Once again there are many to choose from, so take your time and choose one that best suits your needs. As with your footwear, check out the rucksack while carrying a good load prior to leaving for your place of employment. Check whether it is causing any pressure spots, soreness, or irritation, and adjust accordingly. If need be, add padding. The backpack's center of gravity should be high on your back, and its weight should be distributed between the shoulders and the hips 60/40 respectively. This way, your legs will help to bear the weight, and your back will not get strained.

Packing a backpack is also an important skill and one that should be learned before you depart. The most important aspect is deciding what is essential and what is nonessential. Pack a backpack well, and not only should you have what you need, when you need it, but you will also have greater comfort while carrying it.

Weapons

War zones are full of weapons, and Afghanistan is no different, especially as it shares a border with Pakistan, which has a very vibrant arms industry. There are an estimated 20 million illegal arms in circulation in Pakistan. This includes some serious firepower from long-range sniper rifles and heavy machine-guns to RPGs. Many of these weapons have found their way into the hands of the Taliban and onto the weapons market in Afghanistan. You simply need to go to any of the market places in Kabul, Kandahar, Herat, Lashkar Gah, or Mazar-i-Sharif to purchase the weapon of your choice; they will even let you test-fire it right there on the street.

Barry Davies

There are always lots of weapons in a war zone; most local markets sell them openly. A good weapon will cost you only around $50.

There are two things you need to know about weapons: First, how to handle them safely. Second, what kind of weapons you are going to need. It is worth your while investigating all the types of weapons that you may be asked to use. Modern weaponry is vicious; know what you're up against and what you have at your disposal. Trust me, when you have to fight, firepower is everything.

If you carry a gun, then you must be prepared to use it. In the role of a protector, one is often faced with opponents who are happily prepared to die, due to either religious beliefs or the fact that they are off their head on drugs. Either way, they have no fear of death. So when it comes to shooting, don't try anything fancy; just shoot them in the head. Make sure they go down and stay down.

The type of weapons that your PMC will have acquired will very much depend on what is realistically available. While some of the larger companies may have licensed weapons, the chances are you will end up with an odd ball of a firearm or, more than likely, an AK-47 of some description.

Barry Davies

Shoulder rig is excellent for close protection without looking too gung-ho.

Shoulder Holster Rig

If you are employed to provide close protection, you might need to dress fairly smart so as not to embarrass your Principal. In a war zone, the weapon you carry will be either an M4 or a short submachine gun, as this facilities your getting in and out of cars or helicopters with ease. You should also consider wearing a shoulder-mounted rig to carry your weapon when not in use.

Designed mainly for covert use, the shoulder holster rig consists of a shoulder holster and double magazine carrier used to conceal a 9 mm semiautomatic. They are manufactured in water-resistant soft leather or hard Cordura, as the harness is designed to be worn for long periods with maximum comfort. The holster and magazine carrier are fitted with loops for securing to the wearer's belt. The weapon can either be completely withdrawn or fired while still attached to the rig. The British SAS favor such holsters when engaged in bodyguard work and VIP protection.

Lance Cpl. Daniel A. Wulz

Learn how to handle the AK-47; its one of the easiest to acquire.

Kalashnikov (AK-47)

Without a doubt, the AK-47 is the most popular weapon ever made. Millions have been produced in a variety of forms by at least twenty different countries, and it is still used today by the armies of over fifty nations. The country of Mozambique even has a picture of one on its national flag.

Mikhail Kalashnikov's original design, though influenced by the German assault rifles that appeared towards the close of World War II, became a triumph of military practicality. Ease of mass production and incredible reliability made it a real winner. Since the first version, which appeared in the early 1950s, the AK-47 achieved widespread notoriety, and the design was constantly reinvented into various versions around the world.

The AK-47 chambers the short Russian 7.6 2mm round from a 30-round curved magazine and has a cyclic rate of fire close to 600 rpm. The effective range of the AK-47 is about 500 to 1,000 yd, but close up it is a deadly weapon. You can learn more about the AK-47, as it has its own website.*

United States Military

M4

M4

The M4 derives from the original Colt AR15/16, a weapon I used during the Oman War. To be honest, the original AR15 was rubbish and was subject to frequent stoppages—the same cannot be said about the M4. It was the U.S. Special Operations Command

---
* http://kalashnikov.guns.ru/.

(SOCOM) that spotted the need for a short-barreled universal weapon for all of the Special Operations community. The modification, which originates from the M16A2, is fitted with an integral Picatinny-type accessory rail, allowing for several attachments to be fitted. These include various sights, such as the ACOG 4X telescopic, laser pointers (visible and infrared), detachable sound suppressor (silencer), and a modified M203 40 mm grenade launcher. You can even attach a front grip. The M4 carbine is currently used by the U.S. Army, Marine Corps, and SOCOM operators around the world, including Afghanistan.

The M4 is a gas-operated rotating bolt which chambers a 5.56 mm caliber round. It is 757 mm long with the stock folded and 838 mm with it extended; the total weight with a 30 round magazine is 3 kg. It has a cyclic rate of 700–950 rounds per minute and an effective range of 360 m (394 yd).

Keep Your Weapons Clean

No matter what type of weapon you choose, you must keep it clean. Any dirt that gets into it can easily jam the mechanism for when you most need it. A boot brush will get rid of most of the dirt, and you can follow this up by using a tin of canned air, the type used for cleaning the keyboard on laptops, as this will blow out the dirt from all those little places you can't reach. Also, find or make yourself a "pull through" for your weapons so you can clean out the barrel.

You should then wipe your weapon down with light oil for protection, but make sure that it does not get too oily. I personally use an all-purpose soft wipe that is non-toxic and non-abrasive. Take a couple of packs with you, or buy them at the local PX or NAAFI, as they normally stock them.

Finally, don't forget the magazine: Empty the rounds out of your magazines at every opportunity when back in base. Then let the magazine spring relax for a few minutes before reloading the rounds.

9

KNOW WHEN TO GET OUT (AND KNOW WHEN TO RUN)

If you know the enemy and know yourself, you need not fear the result of a hundred battles. If you know yourself but not the enemy, for every victory gained you will also suffer a defeat. If you know neither the enemy nor yourself, you will succumb in every battle.

—Sun Tzu, *The Art of War*

The above quote is the truth of your survival, should you get caught alone and isolated in the middle of a war zone. How well you know the enemy and what are they capable of, as well as the confidence you have in your own skills (plus a whole lot of luck), will help get you to safety. You need to be alert at all times, even at the point when leaving the country of your employment: Don't get complacent until you are in the air, on the ship, or over the border, and have a drink in your hand.

Even before you leave home for your country of employment, you should be thinking about extracting yourself in case of any emergency. Much of this information will be provided by the PMC who are employing you, but ask if it is not forthcoming. If you are traveling into a war zone by yourself, you will need to know who is picking you up, what he or she looks look like, and a mobile phone number to reach that person in case of an emergency. If you are not getting picked up, then plan how you are going to get from the airport to your final destination. Make sure the local taxis are safe.

Unless it is operated by the military, an airport in a war zone is not a very organized place. Even when it is run by the military, do not expect it to be anything like JFK or Heathrow; your baggage can easily go missing. In some airports, such as Libya's, customs employees will confiscate anything they do not like, which includes your body armor (you are allowed to take one set for personal protection). In this case, make sure you have a receipt, and do not leave your bag until you get one. Stand your ground on this one or they will rob you blind . . . the same goes for some of the airports in Nigeria.

Make out a list of useful contact numbers. Most people have them on their phone, but phones get stolen. Copy the list several times and have the copies stashed in every bit of luggage, especially your bug-out bag. The one thing you need to do when running for your life is to have the ability to inform someone else; no matter where he or she is in the world, give that person your location, predicament, and plan. The list should contain some of the following:

- The name and location of your hotel
- The name of the local hospital
- Names and contact information of colleagues
- Contact information of next of kin
- Operation room
- Any number that will be of use if you are alone and isolated

You will also need to know the procedure for a MedEvac, should you get wounded and need shipping to a hospital or even back home. . . . And you need to be sure in advance who pays the bill, as it can be very expensive. Another necessity is the number for your local embassy and the name of any useful contact therein. But trust me, most embassies are not too reliable and tend to be of very little help. You should also ask your employer what plans the company have for a serious situation bug-out, such as:

- Is there an airfield and organized aircraft pickup?
- What are your alternative routes out (road, railway, airport, or sea)?
- What emergency rendezvous have been made?
- Does the company have a safe house?

Author's Note: During SAS "black ops," it was a rule that all labels were detached from clothing, and tattoos had to be removed. This simply meant that if captured, you gave the enemy no starting place to identify you. Religion plays a very large part in most wars, so I would suggest you remove items of religious significance too. This does not mean you should abandon your religion.

Emergency Contacts

Your list of emergency contacts should be informed of your pending employment long before you depart. Always have one person as a main contact—maybe someone from work or college, a desk operator, best friend, or senor family member. Always prep a text message ready to transmit, and update this on a weekly basis or when you move locations. Ensure you always have good communications equipment: one that works. (Don't rely on satellite mobile phones. While they are very good, they also break down at the most critical moment.) Take a variety of service provider SIM cards with you to make sure you can connect. Do not forget that communications do not just have to be from a mobile: email and VHF two-way radios can also help you in a fix.

Alert your prime contact should you be about to undertake a dangerous task or journey, and give him or her a rough time as to when your task should be complete. Have a series of cut-off times, and arrange schedules to make contact: This could simply be a text from you saying, "Bob OK," and a reply from them saying, "Roger that." This will not only make sure your communications are working, it will also give you (and your contacts) peace of mind.

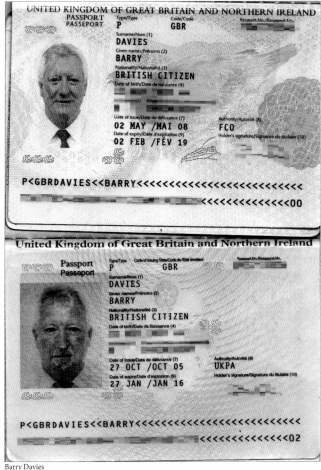

Barry Davies
Always carry two valid passports if you can legally acquire them.

Passports and Visas

Always carry your passport securely on your person, as it lets people know who you are and that you are telling the truth. Check your passport for entry/exit visas to Israel, as many countries will not let you in with an Israeli stamp, which could also sign your death warrant if captured. Equally, if you are travelling to Israel, be careful about what stamps you have in your passport.

Make sure you have a valid visa for the country you are employed in before you leave; even if the country offers visa on arrival, as it can take you ages for you to clear immigration, and rates can sometime be extortionate. If the country you are employed in is still hostile (i.e., no peacekeeping force on hand), then I would suggest you get extra visas for neighboring countries, just in case you need to move quickly.

Money and Credit Cards

As well as having local currency, you should also have enough dollars or Euros to get you out of any emergency. Distribute any cash around your body, and do not flash money around when paying for a lift. Get into the habit of carrying two wallets: In the second one keep some old credit cards and a bit of cheap foreign money to make it look real.

Back your cash up with a couple of good credit cards; make sure your credit cards are up to date, have sufficient value assigned, and are valid. Inform your bank that you are traveling and to where. Using a credit card in a country that the bank has not been told of will trigger a block, which means you then have all the aggravation to get it lifted.

Team Members

If you are working as part of a team, which is fairly normal in PMCs, unless you are known to the other members already, you may find a change of values. Living accommodation in PMCs can vary from

A. R. Howell
Getting on with your fellow team members will make the tour of duty go much smoother.

the plush down to sleeping on a camp bed with a dozen other guys. If you have not been in the forces, this may well seem strange to you; my advice is to give it a little time after you arrive to soak up the lay of the land and see how things are done. Keep your mouth firmly shut, and listen to the conversations. No doubt, as a newcomer you will be pulled into the conversation; some of the members will be downright nosey, some will be curious, and other will ignore you all together.

The workload can be dangerous and hard; many soldiers of fortune have their own way of dealing with the daily routine. Some will act like robots and work, eat, and sleep day in and day out. Others will be euphoric: laughing and joking as they return safely back to base. What you will be experiencing is how each member of the team copes with the stress. They are not being insensitive and hard, they are just expressing their individual coping mechanism. Respect it.

You may also find that your employer, in addition to yourself and fellow American or European members, has employed locals or other nationalities. While you may not share the same room, you may well be sharing the same building. Pay respect to their religious calling, and do not make fun of their unique ways. In essence, accept the conditions, and learn to live with everyone, even those you do not like.

You may also find that the PMC has employed female operators, and you will soon learn that these are equal (or better in some cases) to their male counterparts. However, it is my experience that you should still acknowledge their gender and provide a small amount of extra privacy if they are sharing a room with a lot of men. If you are not sure how to handle it, let the female operators take the initiative; they will soon show you the boundaries.

In every group, there are always jokers and intimidators but, in the main, most of your team members will be welcoming and helpful. Listen to their advice, and learn from them. Establish equality in the team, make sure that the workload is shared out, and that you do your share to the best of your ability.

Secure your personal gear as best as possible, especially your weapon, belt kit, and body armor. If you have a separate armory, store your weapon and gear there; make sure someone always sleeps in the armory if at all possible, and avoid leaving it unattended.

Author's Note: Regular army personnel will often find themselves in an open barn-type building with nothing but a few camp beds. If this happens to you, do not whine or complain about the conditions—make the best of the situation. The first thing you should do is grab yourself a space large enough for you, your bed, and your kit. Look around for any box or crate, which will serve as a table and on which you can keep stuff off the dusty floor. If you are in one location for a long time, you will soon learn how to make your location comfortable, and trust me, you always need a place to come home to where you can shower, clean up, and rest.

Once you get to know the other members and have made friends, discuss with them what they would do if they had to bug out quickly and how have they prepared themselves for any emergency. You might have some good ideas, but it is always best to listen to everyone, as you will always learn something. If the local police force have been retrained or considered as safe, make a list of the various police stations across the county. However, if they are part of the problem—and many are—give them a wide berth.

Individual Room

Should you find the sort of good accommodation where each team member is allocated an individual room, make it as secure as possible. Check that all windows are locked, and use curtains if exposed to the rooftops of adjacent buildings. Hide your valuables as best as possible for while you are out, and lock the door at all times. If the door has no lock, then use a wedge at night to prevent, or at least delay, anyone from entering your room. The crime rate is very high in any war zone, and if your company employs housekeepers and kitchen staff, you should not be surprised if your iPod goes missing.

Safe House

In places such as Mogadishu and parts of northern Nigeria, it is always worth having a "safe house." This can be a single room, office, or a whole building, depending on your needs. Whatever the size, it should offer the best protection possible. Safe house means just that: It is safe (at least for a stated time) and offers shelter, accommodations, food, and water.

In a war zone, it can be extremely difficult to locate a safe house; stock it with necessities, and hope that it remains undetected. One PMC working in Africa dug out several underground hides and equipped these with food, water, and medical equipment. The exact location was known only to members of the PMC and kept secret: These sites were used by operators several times.

Cache Locations

The British SAS has, for many years, used cache locations to assist in resupply or to aid its members on the run from the enemy. The idea of a cache is very simple: It is basically a type of time capsule that is buried and contains all the items a soldier will need to help him or her stay alive: food, water, navigation equipment, etc. They are sometimes made up of surplus equipment and stores that are no longer required and, to save transport, they are sealed in plastic bags and then put in old ammunition boxes before being buried and a full "cache report" being drawn up. At other times (mainly during the cold war), purpose-made cache containers have been used, but these are not as popular as they once were.

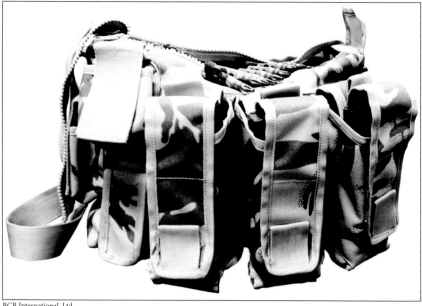

BCB International, Ltd.

Combat ammo bag.

Personal Combat Bag

Everyone who works in a war zone, from regular soldiers to journalists, will tell you that you need a "grab bag" of some description.

While this is essentially a bag that can be grabbed in a hurry, it also serves the purpose of a simple support bag on a day-to-day basis. The idea of a personal combat bag actually grew out of the PMC industry and in some cases has been adopted by the regular military. The difference between the combat bag and a grab bag is the former's ability to carry extra ammunition, magazines, and grenades. However, I would also advocate adding the following equipment:

- A good method of communication: mobile, satellite, or two-way VHS radio.
- SatNav or compass and escape map with safe locations (hide these as best you can).
- Medical pack for dealing with extreme trauma, such as wounds from IEDs.
- Torch and battery, a radio with a hand crank, and the ability to recharge your cell phone battery.
- Dried food, such as boiled sweets, dried fruit, and OXO cubes (for flavor and salt content).
- Compressed sleeping bag: A large poly bag will do in an emergency.
- Half a roll of toilet paper: lightweight, and it will keep you sane.
- Plus anything else that will help you in your particular location and situation.
- Add a couple of packs of cigarettes (to bribe even if you do not smoke).
- Simple things such as a strip of purification tablets, duct tape, etc.

Water should be always with you; purchase a CamelBak hydration system or something similar. Every soldier has one, and every soldier uses his or her hydration system to keep his or her body operational. If you find yourself in an emergency situation where you are running for your life, your hydration system could mean the difference between survival and dying.

Author's Note: A friend of mine who works in Africa always carries a Red Cross jacket, but he is a trained medic, so he is able to get away with it. If captured, he will play the role of a Red Cross medic: For some reason best known to themselves, rebels and insurgents are less likely to kill or harm a recognized medic. The same goes for cooks, as few cooks are killed in a war zone. However, if asked to provide medical assistance to a local Islamic woman, no matter what the urgency, make sure you ask permission of the men around you to touch her before you proceed.

A. R. Howell

Only approach help when you are 100 percent sure it is safe to do so, and don't flash your money around.

Approaching Help

If you should find yourself alone or with others and running for safety, you will almost certainly come across the possibility of people who could help you. I say "could," because you are equally prone to surrendering yourself to the enemy. In my personal experience, you are best avoiding or approaching possible help until you have put lots of distance between yourself and the location of the incident that forced you to flee in the first instance.

This is where local knowledge is vital to your continued existence and survival.

War zones, for the most part, are ungoverned and lawless, and you are best treating everyone as hostile. There is no law in war, other than the law of the gun. That said, most civilians, if treated with respect and no signs of aggression, will help you with food and water; whether or not they inform the local insurgents of your presence will depend on the individuals. Remember that not all civilians are rebels or bad guys, and not all those in uniform are good.

If you know the location of a Police station that is friendly or a military compound with foreign troops, you should make your way to these locations as quickly as possible. Hospitals on the other hand (unless known to be friendly) should be avoided, as they will almost always contain civilians and possible rebels who have been wounded, together with their families. There will be a lot of hatred for any foreigner.

Your escape plan needs to have a number of safe locations for you to head towards. You should know at least two of them by heart and you should know what stands between you and the place of safety you are trying to reach. If you are out in the open desert, you can travel at night with the darkness to protect you. If you are in an urban area, you should camouflage your identity as best as possible and move during the day (late afternoon during siesta time). Do not move through an urban area during the hours of darkness, as this is one quick way to get yourself killed. The idea is to blend in with the locals: Do not shave; instead, grow a beard. Leave your hair grow long, and have it cut in a local style if there is one. Wear clothing similar to that of the local population and head gear if applicable.

Do not become complacent. The longer you stay and work in a war zone, the more you will become accustomed to everyday violence. You will start to relax because of the "nothing has ever happened to me" syndrome. This is a dangerous state of mind, and one that will eventually get you killed. Most locals will wear sandals, but trainers are becoming more fashionable, so get a good pair but make them look old and dirty. If you have a weapon such as an AK-47, you will fit in nicely in many hostile areas.

- Carry any weapon in the same fashion as the locals.
- Hide your combat bag as best as you can.
- Move at the same pace as the locals.
- Do NOT wear sunglasses.
- Keep your head down, and don't look locals in the eye.
- If trouble starts around, you just get out of the way. DO NOT turn and run.
- Stay cool and don't panic.

Government Agencies

In any war zone, you will always find a number of government agency employees lurking about. While some are quite good and profes-sional, others will stab you in the back at the first opportunity. It is not uncommon for government agencies to approach soldiers of for-tune and ask them to do a little research on their behalf, although they normally select personnel from commercial firms. Agencies also work very closely with Special Forces, and it may prove advantageous to build up a relationship. However, unless approached, I would suggest you stay well clear of anyone claiming to be from a government agency.

Some countries that PMCs become involved with can be very dangerous, and while the PMC has a legitimate right to work in the country, it does not always fit in with certain governments who might have an agenda of their own. The CIA, MI6, plus many others may well have an interest in the country you are operating in. If there is a clash of ideals, you could find yourself being fingered by the agencies. This happened to Simon Mann.

Simon Mann in court when he was with the SAS.

Author's Note: I knew Simon Mann when he served with "G" Squadron SAS; he was a good officer, but not one that stood out for anything special. So it came as a surprise when I heard he had been arrested in Zimbabwe. He had been involved in a coup to bring down the dictatorship that ruled Equatorial Guinea in 2004. The coup was backed by wealthy right-wing businessmen in Britain and South Africa. Mann and his team of around seventy men planned to topple the president, Teodora Obiang Nguema Mbasogo, and replace him with a puppet president, Severo Moto. In return, Mann and his backers would then lay claim to the lucrative oil fields and security contracts. The CIA tipped off the Angolan intelligence service in order to have the coup fail, and Mann and his men were arrested.

Simon Mann was given a long prison sentence (thirty-four years) and sent to Black Beach prison in the capital of Malabo, a prison where conditions were described by Amnesty International as "life threatening." Starvation, nonexistent medical facilities, and torture were all commonplace.

Simon Mann was released on November 2, 2009, on humanitarian grounds and returned to the United Kingdom. Strangely enough, he has now gone back to Equatorial Guinea to work for the president, although the role and type of work is undisclosed.

Barry Davies

Simon Mann while with the SAS.

Traveling through Hostile Territory

When working as a soldier of fortune in a war zone, it is not beyond possibility that you will find yourself alone or with others in an emergency situation. Your convoy may have been ambushed, your vehicles destroyed, forcing you to go on foot, or you've become separated. It could also be that the war does not go well and your location is in danger of being overrun, forcing you to move.

No matter how your emergency arises, you may find yourself in a hostile predicament in which you are forced to evade the foe; your one great hope is one of rescue or getting yourself to a known safe area. In the short term, your survival will depend on where you are and what kind of situation you are in; it will also depend on your ability to contact friendly forces who will be able to form a rescue. Then it will be a case of keeping yourself undetected and alive until rescue arrives. When people are looking for you, and in many cases risking their lives in the process, it is up to you to help by providing the best situation report possible. Always make sure you have your mobile charged up and handy and that your phone's GPS is activated.

Author's Note: All the years I served in the SAS, I carried an escape and survival kit. The need for such a kit is not so obvious today, but if I were working as a soldier of fortune, I would certainly carry a few tools in addition to those in my combat bag, most of which I would hide in my clothing. They include:

- Lock picking tools
- Scalpel blade
- Button compass
- Escape map of the area in which I was operating

Don't forget, if you are kidnapped, items such as your combat bag, watch, mobile phone, SatNav, etc., will be taken from you immediately. Even when your captors strip you to the bone and leave you only your basic clothing, remember the one thing they cannot take from you is your brain. Even with nothing, it is possible to navigate, remember where safe locations are, and improvise. While the subject is far too vast for this book, I have detailed a simple trick below.

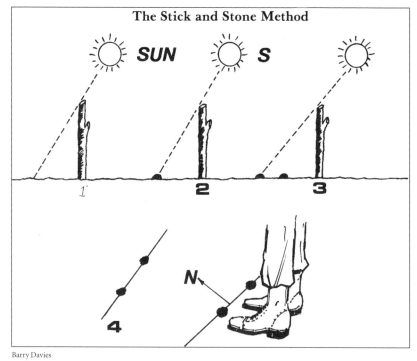

Barry Davies

Stick and stone method of finding direction.

If you are in hostile territory, it would be wise to avoid all contact with the local population and to remain unseen. This may involve travelling only under cover of darkness or using a disguise to travel under. Do not take chances. Stay alert. Any successful evasion will rely on the following:

- Preparation prior to leaving your base or safe location.
- Having a workable standard operating procedure (SOP).
- Preparing yourself both mentally and physically; getting a grip on the situation.
- Deciding on a location to head for and the direction and route you will travel.
- Covering as many of the "what ifs" as you can and considering all the things that could go wrong.
- In a safe place, checking the contacts on the list in your combat bag and making contact as soon as it is safe to do so.

Stick and Stone Method for Finding Direction

With nothing more than a stick and two small stones, it is possible to get a good direction indicator, providing the sun is shining. Cut any stick or twig about 1 m (1 yd) long, and push it upright into the ground. It is best to choose a very level spot where a clear shadow will be cast by the sun. Mark the very end of the shadow as accurately as possible, using a small pebble or stone. Wait for 15–20 minutes, and you will observe that the shadow has moved. Again, mark the tip of this shadow with another pebble or stone. Draw a straight line on the earth, running through the first stone and on through the second: This line will run from west to east.

Now stand with the toe of your LEFT foot against the line and close to the FIRST stone. Bring the toe of your RIGHT foot to the line near to the SECOND stone. YOU ARE NOW FACING NORTH.

The accuracy of your direction finding will depend on the accuracy with which the shadow tips are marked, and the care you take in placing your toes to the line. A line drawn at right angles to your east/west line will produce a north/south indicator. From these cardinal points, it will be a simple matter to calculate any other desired direction.

Summary

There is one last thing to learn, and it is probably the most important: Know when to get out. When you arrive in your theater of operation, check out your escape routes. The best way to do this is to ask another member of your company or another PMC what he or she intends to do. You will need to know the following basics:

- The nearest safe location and also where the "green area" is (if there is one).
- Who controls the airport and/or ports?
- Which roads lead to safety and which lead into enemy-held territory?
- What are the nearest neighboring friendly countries?
- Escape transportation: Where is it located, and is it kept fully fuelled?

Above all, keep your eyes and ears open for local intelligence, and make a note of it. Monitor how the war is going and whether or not the opposition is gaining ground. If they are winning, give yourself a minimum barrier of time and distance for when you should extract yourself. Remember that as a soldier of fortune, you have little or no chance of survival if captured; at best you will spend years in a crap prison, and at worst your head will be removed. Use this thought as a motivation to act promptly and stay alive.

10

PMC MEDIC

Most people will come to this chapter and simply brush past it because they are not interested in learning about medical topics, that it is for other people. Well I have to tell you, if you are going to work for any PMC, you will need at least some basic grounding in first aid and basic medical procedures. . . . If not for yourself, then for the

IED wounds are horrific and generally mean loss of both limbs.

benefit of your fellow operators. I have cut this chapter to the bone, but my advice is to read it and continue to study if you intend to be a soldier of fortune.

Why is basic medical knowledge so important? Well the one thing about being a PMC employee is the frequent hazard of getting shot or blown to pieces by an IED (see Annex C for a few examples). Additionally, a war zone is full of other hazards, such as mosquitoes, rabid dogs, and flesh-feeding flies—all of which carry some deadly disease. The British SAS has a medic for every four men, as do many other Special Forces; the medic's aim is to provide emergency medical cover and keep the patient alive until an evac can be arranged. A good Special Forces medic is capable of dealing with most traumatic emergencies, to the point of doing serious surgery.

BCB International, Ltd.

You and your team are going to need a good medical pack.

Medical Pack

Knowledge of even basic first aid skills is a useful and valuable accomplishment in everyday life, but when you are in a war zone, these skills take on immeasurable importance. Even when medical training and equipment is limited—or totally nonexistent—it is always possible to save a life if the priorities of first aid are administered. It is important to put together a small emergency medical kit based on your own personal medical skills. Obviously, if you are not trained as a medic, you should only include basic items, as detailed below:

- **Plasters:** Carry various sizes and shapes of waterproof plasters. Larger plasters are best as they can always be cut down if necessary. Keep your plasters together in a waterproof sachet. If you are unable to administer stitches, butterfly sutures will prove successful in closing small wounds.

- **Large Wound Dressings:** You should carry at least two large wound dressings in your medical kit. As any soldier will tell you, always have them ready for immediate use.

- **Surgical Blades:** Two surgical blades take up little space and are best left in their protective sterile wrapping.

- **Pain Control:** Pain control is vital to the recovery of a patient, and your medic pack should contain a wide range of pain-relieving drugs, both prescription and nonprescription. These include aspirin, acetaminophen, co-dydramol, dihydrocodeine, DF118, and morphine (if you can legally get it). Carry a strip of about twenty-eight tablets for each kind.

- **Antibiotics:** You should have at least one or two good antibiotics in your pack; you should also know when to administer them and for what condition. Amoxicillin or a broad-spectrum antibiotic such as tetracycline will suffice. Once again, these are prescription drugs, unless you happen to be in a country where they are openly sold over the counter.

- **Mosquito Repellent:** The chances of contracting malaria and other mosquito-borne diseases can be reduced if the correct precautions are taken. Anti-malarial tablets, as prescribed by a doctor and need to be taken, but it is just as important to deter the insects from biting you in the first place. Therefore, it is recommended that you include a mosquito/insect repellent in your kit.

- **Antihistamine Cream:** Bites and itches can be distracting to your work, especially if you are guarding a Principal. Antihistamine cream will soothe the severe itching and irritation that insect bites or allergies can cause. Antihistamine tablets can be carried as an alternative, but be aware that some cause drowsiness.

- **All-Purpose Antiseptic:** Select a good strong all-purpose antiseptic, and carry at least two tubes. Small cuts can easily get infected in unhygienic working conditions such as a war zone.

- **Salt:** Salt is essential when travelling in tropical or hot climates. Carry a small amount to make sure that the salt balance in the body is maintained. Try to reserve this resource for medical uses only and refrain from using it for culinary purposes.

Author's Note: As an individual, you may also consider carrying pain relievers and anti-diarrhea medicines on your person if you are operating away from the main medic pack. I would recommend that in addition to the main medical pack you have a small one on your person at all times.

Life-Support Pack

Most current theaters of war present a very high risk of IED injuries; so in addition to your medical pack, I would also suggest a "life-support pack." The equipment your pack holds should be geared to deal with severe wounds and injuries, including those caused by blast and fragmentation. Resuscitation, ventilation, and aspiration equipment; intubation equipment; and intravenous administration kits should be included in the pack, as well as dressings, tracheotomy and cricothyrotomy, burn treatment, and limb immobilization equipment—basically everything you need to treat a casualty who has just lost his two legs and left arm.

Priorities of Treatment

After any incident which has produced casualties, the first task is to establish a process of prioritization of the wounded. Those who require urgent assistance to prevent immediate death (those suffering from asphyxia) must be given priority. Shock caused by major body loss and severe hemorrhaging must also be assessed. The task is to identify the injury and how long the patient will live without assistance. Also, if assistance is given, will it prove beneficial? While formulating your priorities, keep in mind these rules:

- Do not panic, no matter how serious the situation may look. Panic means that you will think less clearly and therefore waste time or make the wrong decisions.

- Individual casualties will need to be assessed as to their injuries. For this you will need to use all your senses: ask (if the casualty is conscious), look (and, if possible, feel over the body for broken bones and blood), listen, smell, think, and then act. If the casualties are conscious, ask them to describe their symptoms and what they feel may be wrong with them.

- Exclude taking any action that will put you in danger, especially if the incident is IED related. If you become injured, then you will be in no position to help anyone else.

- Once you have considered your actions first, act quickly and carefully. Boost the morale of the casualties. Offer comfort, reassurance, and encouragement, thus building their mental strength to live.

United States Military

Use simple resuscitation method to keep people breathing.

Breathing

If we stop breathing even for even a few minutes, our brain will start to sustain damage. The longer we go without air, the greater the damage. If a casualty is unconscious, choking, or having trouble breathing, then he or she must be treated urgently. In the case of an unconscious casualty, check for breathing and a pulse. If one or neither is present, then emergency treatment must be given immediately. Urgent assistance must also be given to anyone with choking or breathing difficulties.

Check a casualty's breathing by placing your ear close to his or her nose and mouth and looking down over the person's chest and abdomen. If he or she is breathing, you should be able to feel and hear the air as well as see chest and abdominal movement. If these signs of breathing are absent, immediate action must be taken.

First make sure that the airway is clear:

Gently tilt the casualty's head back while lifting the chin with the other hand. Doing this will automatically open the airway and will also lift the tongue from the back of the throat so that it will not cause an obstruction. Support the head in the tilted position with a hand on the forehead, and check inside the mouth for any object or substance that may be causing a blockage (dentures, vomit). If any of these are present, gently remove the blockage without touching the back of the throat, as this may cause a swelling of the throat tissues. In many cases, these actions may just be enough to enable the casualty to breathe again. If this is the case and the person also has a pulse, place him or her into the recovery position and make sure he or she is checked periodically. Many people will say it is not safe to move an injured person, and any visible injury to the front or back of the head may also indicate that the casualty has damaged his or her neck or spine. In this case, maintaining an open airway will still be a priority over other injuries. However, it is recommended that a collar or head support be improvised in order to keep the head properly positioned.

If the casualty is still not breathing, then extra steps must be taken to ensure that the person gets some oxygen into his or her body. This used to be achieved through mouth-to-mouth resuscitation; however, a new a technique, which uses only chest compression, is currently preferred. Place the casualty flat on his or her back on a firm surface. Kneel beside the person and locate the bottom of his or her breastbone: This is found where the bottom two ribs meet. Place the heel of one hand about three fingers' width up from this point, and place your other hand on top and interlock the fingers. Lean forward over the casualty so that your weight is pressing vertically on the casualty's chest, and perform compressions to the musical tune of "Staying Alive." The song has exactly the right tempo of 100 chest compressions per minute. Continue with this

United States Military

Use direct pressure to stop bleeding.

cycle for a minute before checking on heartbeat and breathing. If neither is present, continue with the chest compressions either until the casualty's heartbeat is restored, help arrives, or you become too exhausted to continue.

Bleeding

Once breathing and circulation are restored, the next priority is the bleeding. Bleeding may be external or internal. Internal bleeding is almost impossible to treat with first aid, but external bleeding can be controlled. Wounds present two main problems: Firstly, extensive bleeding can cause shock to develop and will, if not controlled, lead to death. Secondly, any break in the skin will let infection in, so it is imperative that the wound site is kept as clean as possible. There are three procedures with which to stop the bleeding.

Direct Pressure

Use a sterile dressing if you have one; if not, find a clean piece of cloth. Place the dressing on the wound, and press on it gently but firmly. If you have no dressing available, then you may have to use your hand, but bear in mind the dangers of infecting the wound. Use only dressings that are large enough to cover both the wound and part of the surrounding area. It is possible that the first dressing will become soaked with blood. If this happens, lay a second dressing over the first and, if necessary, a third over the second. By tying a bandage around the wound and dressings, you will be able to keep the dressings in place with a continued firm pressure. It is important though that the bandage is not tied too tight, as this will restrict the flow of blood to the whole area.

Some large wounds will tend to gape. If you have suitable dressings, you may use these to bring the edges of the wound together; otherwise you may have to use your hand. Blood flow from a large wound may be stopped by applying firm pressure, preferably with a pad of dressings, to where the bleeding is the greatest. Using pressure on the wound helps the body's own mechanisms to slow down and finally stop the bleeding.

The elevation of an injured limb will not only reduce the flow of blood to the damaged area, but also helps the veins to drain blood away. This helps with the reduction of blood loss through the wound. The limb should be above the heart level, as long as it is comfortable for the casualty and not liable to make any other injury worse.

Indirect Pressure

If due to the severity of the bleeding the techniques described in the previous section do not work, then indirect pressure should be tried. However, this only works on arterial bleeding, so it is important to identify what type of bleeding you are dealing with.

Arterial bleeding takes place from vessels that are carrying filtered and oxygenated blood away from the heart and lungs. It has no impurities and is therefore bright red. It will also spurt out of the wound in time with the heartbeat.

Venous bleeding takes place from vessels that are carrying blood full of impurities away from the tissues and to the heart and lungs to be filtered and re-oxygenated. As venous blood is low in oxygen, it is dark red in color. It runs steadily or gushes from a wound at a steady rate.

- Indirect pressure works by using pressure points. These are found where an artery crosses a bone near to the surface of the skin. For survival purposes it is best to concentrate on the four points that flow to each limb.

- The pressure points in the arms are found down the center of the inner side of the upper arm, on the brachial arteries.

- The main leg artery is the femoral artery, and it is located down the inside of the thigh. The pressure point for this artery is in the middle of the "bikini line," or crease, where the leg meets the body. It is often easier to locate if the knee is bent so as to create the groin crease. Press firmly at this point against the bones of the pelvis.

- Locate the pressure point and, placing the thumb or fingers on it, apply enough pressure so as to flatten the artery against the bone. This should stop the blood flow.

- Pressure must not be kept on for any longer than ten minutes or else other healthy tissue will be damaged through lack of blood. While using indirect pressure, the wound may be dressed more effectively; do not, however, use a tourniquet as this may cause tissue damage.

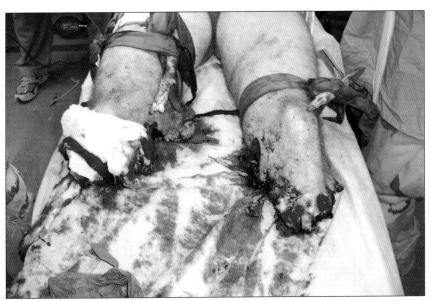

United States Military
For any IED explosion, you will almost certainly need a tourniquet.

Tourniquet

If after an IED explosion the damage to a limb is so severe, or there has been a blast amputation and part of the limb is missing, direct pressure will not stop the bleeding, and you will need to apply a tourniquet. The tourniquet can be made from whatever cloth is at hand, but avoid any thin material that will cut into the flesh. Place it around the extremity, between the wound and the heart, 5 to 10 cm (2 to 4 in) above the wound site. Never place it directly over the wound or fracture. Use a stick as a handle to tighten the tourniquet and tighten it only enough to stop blood flow. Clean and bandage the wound. **The tourniquet must be slowly released every 10-15 minutes for a period of 1-2 minutes. However, you should continue to apply direct pressure at all times.** It must be stressed that applying a tourniquet to prevent blood flow is a dangerous procedure and should only be attempted when all else has failed.

Author's Note: Many soldiers operating in Afghanistan have taken up the habit of wearing tourniquets around their upper arms and legs prior to leaving the safety of their compound to go out on patrol. The frequency of IEDs and the horrendous damages caused by stepping on one has required them to undertake this anticipated procedure.

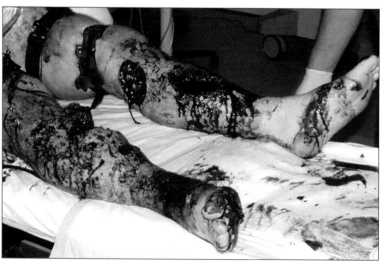

United States Military

Fracture.

Fractures

Fractures normally occur during an accident in which a body has stumbled unrestrained or has been hit by a flying object. The reason for a fracture is fairly evident, and a fracture is followed by a sharp increase in pain when movement of the affected part is attempted. Pronounced swelling, bruising, distortion, and tenderness at the site of the injury are also good indicators of a fracture. An injured limb may look deformed or shortened, and a distinctive grating sound may be heard while attempting to move the limb.

Splints can be used or can be improvised, but make sure the splint is padded and that it supports the joints both above and below the fracture. In the case of a leg fracture, if no suitable substitute for a splint can be found in your environment, immobilize it by tying it to the good leg instead. If possible, elevate and support the fractured limb as this will help to reduce any swelling or any chance of the casualty going into shock. Make sure that the casualty receives plenty of rest.

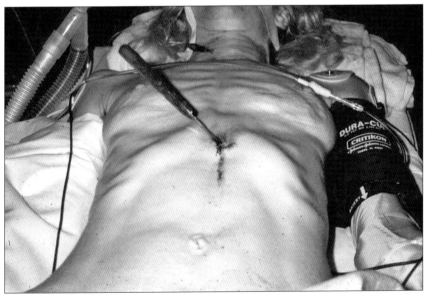

United States Military

Try not to remove the foreign bodies, such as a knife from the body, just pad around it and get the patient to a hospital.

Foreign Bodies

While explosives have a tendency to remove limbs, those not so close to the blast may also receive injury from flying debris. Smaller fragments and projectiles ripped up by the blast can cause serious damage. However, unless life threatening, foreign bodies should be left in place, as pulling at them may do further damage. Instead, control the bleeding by direct pressure, squeezing the wound in line with the foreign body. Next, form a padded ring that fits neatly over the protruding object, and secure it with a dressing.

Sucking Wounds

Gunshot hits to the chest will cause sucking wounds. If air is allowed to enter the lungs from puncture wounds to the chest or back, then a sucking wound will develop. Always check for sucking wounds if missiles of any form have penetrated deeply or a rib is protruding from the chest or back. The lung on the affected side will collapse, and as the casualty breaths in, the sucked air will also impair the efficiency of the good lung if the condition goes untreated. The result is a lack of oxygen reaching the blood stream, which in turn may cause asphyxia. Check for the following:

- Chest pain
- Sound of air being sucked in from the chest
- Difficulty in breathing
- Bright blood bubbling from a chest wound
- Blueness around the mouth

If a sucking wound is suspected, immediately cover the area with your hand. Support the casualty in a lopsided sitting position with the functioning lung uppermost. Cover the wound with a clean dressing, and place a plastic sheet over the top so that the plastic overlaps the dressing and wound. Tape the dressing down to form an airtight seal. If a foreign body is present in the wound, do not attempt removal, but pack with a ring as described above and fit an airtight seal.

Concussion and Skull Fractures

Being in a vehicle when it is hit by an explosive device will result in a concussion or, if the head has been thrown against a metal hull, a skull fracture. Skull fractures and concussions are also common in vehicle convoys in Afghanistan. A concussion is a temporary disturbance of the brain normally due to a severe blow or shaking. If conscious, the casualty should be made to lie down with the head and shoulders supported. If unconscious, make sure that the person is breathing and has a pulse—if not, carry out artificial ventilation and chest compressions immediately. If the casualty is unconscious but breathing and pulse are normal, turn the person into the recovery position and maintain a close check on the vital signs. In either case, make sure that the casualty is kept warm and quiet and handled carefully. Apply a light padding to the injured area, and hold it in place with a dressing. If blood is being discharged from an ear, lightly cover but do not block. A concussion is normally only a temporary disturbance from which the chances of recovery are good. A skull fracture or concussion must be suspected if any or all of the following symptoms are present:

- Obvious head wound, bruise, or soft or depressed area on the scalp
- Unconsciousness, even for a short period of time
- Clear or watery blood coming from the ears or nose
- Blood in the white of the eye
- Unequal or unresponsive pupils of the eyes
- Steady deterioration in responsiveness to external stimuli

Burns

Naked flames, boiling water, electrical devices, friction, acid, liquid oxygen, freezing metal, and the sun all cause skin burns. The severity of the burn and the amount of body area affected will determine the casualty's survival chances. Burns caused by a naked flame should be cooled immediately to limit the damage caused by heat to the

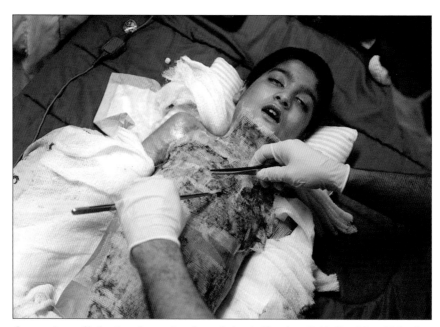

A young boy suffering from horrendous burns is treated in a hospital in Kandahar, Afghanistan.

skin tissues. Either pour cold water slowly over the affected part or immerse it totally in clean cold water. This should be done for at least ten minutes to stop further tissue damage and to reduce pain and swelling. Do not attempt to remove any charred fibers that have stuck to the burn, but remove any restrictive clothing around the site to prevent further swelling. Once the burn has been cooled, a dressing should be immediately applied to limit the possibility of it becoming infected. The dressing should be sterile and made of a non-fluffy material. Avoid adhesive dressings, as these will only aggravate the injury and cause more damage. Do not be tempted to burst any blisters that form, as these provide a protective layer. If polythene bags are available, they can be used to cover the burnt limb and help stop further infection. To reduce the possibility of shock setting in, lay the survivor down and keep him or her warm and comforted. If the casualty is unconscious, turn him or her over into the recovery position and monitor the person's breathing and pulse closely.

- Cool the burnt area by immersion in cold clean water, or fresh snow.
- Protect hands and feet from further infection with a sealed polythene bag.
- Do not use adhesive or fluffy dressing.
- Do not break blisters or remove loose skin.
- Do not apply ointment, oils, or fats to the burn.

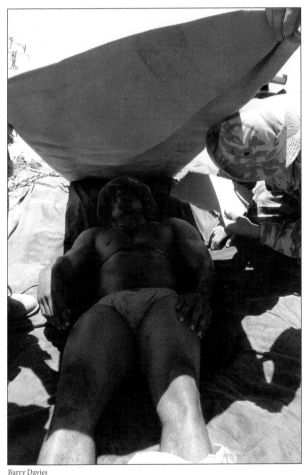

Barry Davies

Heat exhaustion is common and will incapacitate you just a quickly as a bullet hitting you, so stay cool.

Dealing with Heat Exhaustion

Trust me when I say that heat exhaustion is a real killer. It is normally caused by excessive sweating due to hot weather and arduous routines. The simple answer is to rest in the shade, cooling the head with a water-drenched cloth while replenishing body fluids with small mouthfuls of water. Of secondary importance to water is salt. The normal human requires 10 g (1/3 oz) each day to maintain a healthy balance. Sweat contains salt as well as water, and this loss must be replaced. If it is not, then you will suffer from heat stroke, heat exhaustion, and muscular cramps. The signs of salt deficiency are sudden weakness and a hot dry sensation to the body. Resting and a small pinch of salt added to a mug of water will eliminate these feelings very quickly. In dry desert or sweaty jungle conditions, it is advisable to add a small amount of salt to <u>ALL</u> your fluid intake. DO NOT ignore the signs of heat stroke.

Look After Yourself

War zones—apart from the violence—have their fair share of health issues, and it would be wise to consult your doctor ahead of your departure to discuss the health risks. Bring any prescription medicine you may need from your home country, and don't count on being able to find it locally.

Respiratory diseases such as tuberculosis and food-related illness are common, and malaria is a high risk in Libya, Iraq, and Afghanistan. Due to poor sanitation, flies are particularly heinous, especially during the spring. Food should be approached with a discerning eye, as hygiene standards can often be lacking. Bottled water is also advised unless you have your own purification system. Libya, Iraq, and Afghanistan are also some of the dustiest countries in the world, and you should think about some protection for your eyes, nose, and mouth.

Personal Hygiene

Prior to taking up work in any foreign country, you are advised to seek current medical advice, especially on what immunizations and extra precautions you need to take. Many diseases, such as typhoid, paratyphoid, yellow fever, typhus, tetanus, cholera, and hepatitis, can all be vaccinated against.

Proper hygiene, care in preparation of food and drink, waste disposal, insect and rodent control, and immunizations will greatly reduce the causes and number of potential diseases. Prevention of disease during any survival situation cannot be overstated. Personal hygiene is important if you wish to protect yourself against disease and infestations. The sun is a useful agent against disease, as few bacteria or viruses can survive exposure to ultraviolet light.

In my own experience, the two most important items of personal hygiene are your teeth and feet. Keep these clean, and you will avoid most of the problems in a war zone not related to direct conflict.

Teeth should be cleaned at least twice a day: early in the morning and late at night. If a mouthwash is available, use it. If not, improvise with a solution made from salt water. Clean teeth reduce the risk of serious stomach upsets.

Feet require constant maintenance, and blisters or ingrown toenails can be extremely painful and prevent you from walking. Foot blisters are usually caused by ill-fitting boots, poor-quality socks, or loose laces combined with long periods of having to walk over rough, uneven ground. Stop and treat small blisters immediately by covering them with surgical tape. A severe blister is often filled with fluid and can be made more comfortable if the fluid is removed. Large blisters which look about to burst and rupture the skin should be punctured with a sterilized needle and thread. Run the needle through the blister, leaving the threads hanging out, as this will ensure that the fluid drains without creating a large break in the skin. Make sure that the surrounding area is kept thoroughly clean and dry.

Ingrown toenails should be treated as soon as they become apparent. Without removing the nail, the best method is to shave the top center of the nail with a razor blade from your survival kit. Skim the middle third of the nail, shaving from the bed towards the nail-tip. Place a thin piece of plastic under the nail to prevent accidentally cutting the toe front. When the nail is thin enough, it will buckle into a ridge and relieve the outer pressure. Removing the nail altogether should be avoided, as this will require a dressing and may prevent you from walking for several days.

Boots are subject to much abuse in a desert environment: The leather will dry out in the heat and crack unless cared for. Make sure that you remove sand, stones, or insects from your boots: Apart from being uncomfortable, they can cause blisters and wounds. Choose

A. R. Howell

Diarrhea is easy to catch, especially when food preparation is far from ideal. Good hygiene will reduce the risk of diarrhea.

your boots with great care; make sure they fit comfortably and you are happy wearing them for at least twelve hours a day.

Diarrhea

Most who work overseas—especially in a war zone—will suffer from bouts of continuous diarrhea. Although unpleasant, they pose no threat to life, and the disorder is usually self-limiting. Diarrhea will normally develop as a result of consuming contaminated food or water, although malaria, cholera, and salmonella produce a similar disorder. Diarrhea is diagnosed when the number of daily bowel movements is increased by a factor of two or more, with stools being squishy and watery. Water loss must be replaced using sterile water (boiled) mixed with a little salt. Check medical packs for any electrolyte powders and diarrhea medication. If nothing can be found, a small amount of ground charcoal taken from a cold wood-burning fire will settle the stomach when swallowed. Alternatively, a juice made from potassium-rich fruit such as apples and oranges will help, as will honey if it can be found.

Water

Dehydration is a slow process that occurs when bodily fluid is lost at a greater rate than it is being replaced. On average, if no water is consumed, a normal adult will start to deteriorate after four days, with death occurring within the first week. You will need at least 5 L (1.3 gal) of water a day in addition to the other fluids you drink. I would suggest using a water bladder such as a CamelBak to carry your water supply, the size and amount depending on the operation you are doing.

Mosquitoes

Any insect bite has the potential to introduce infection, but tropical mosquitoes are the carriers of several dangerous diseases that can be fatal. Diseases and parasites such as malaria, filariasis, yellow fever, and dengue fever, as well as various forms of encephalitis, are all carried by the mosquito. As mosquitoes breed in stagnant, sluggish water or swampy ground (however, you also find them in dry sunny areas of the world), you would be well advised to avoid camping near any of these, aiming for higher ground where possible. They bite mostly during the late evening and nighttime—although those that carry Dengue fever also bite in the daytime. Use any available insect repellent. Make sure that exposed skin is covered as much as possible by tucking in clothing: trouser legs into socks, and sleeves into gloves. Cover your body with mosquito netting, parachute material, handkerchiefs, or anything else that may be improvised. Smearing mud over any exposed areas of skin will also reduce the amount of mosquito bites.

Rabies

Rabies is still widespread throughout Europe, India, South America, and Africa. Although it can be spread by a variety of infected animals, it is normally transmitted to man by dogs. A single bite—or even a lick—from an infected animal can infect a person. Those infected with rabies almost always die . . . and it is not a pleasant death.

If you intend to travel in an area where the threat of rabies is high, seek advice on a rabies inoculation. This will slow down the speed at which the disease attacks your body—normally around thirty days. If you are bitten, wash the wound with soap and water as quickly as possible, but do not scrub. Remove any saliva from the wound or the surrounding area by placing it under a tap or rinsing it from a water bottle. To confirm or exclude whether someone has been infected, it is vital to examine the infected animal immediately. Without causing any further damage to anyone else, make all efforts to capture or isolate the beast. If you suspect a rabid animal has bitten you, you

Author's Note: In my life, I have seen two people die from rabies. Trust me, it is not a death you want, as you will beg for someone to end your life.

must seek post-prophylactic treatment as soon as possible. Once the symptoms of rabies have appeared, after an inoculation period of between thirty to sixty days, death is often unavoidable.

Dog Attack

While dogs are quite rare in most of the Middle East, they are still abundant throughout the rest of the world. Many are disease ridden and infested with a host of nasty ticks and fleas. If a dog is charging at you, try to break its momentum. This can be achieved by standing exposed next to the corner of a vehicle or building. Wait until the dog is a few feet away, then at the last second move rapidly around the corner, but facing the dog's charge. The dog is forced to slow in order to turn; take advantage of this.

Shoot the dog if you are armed; if you have no weapon, take off your coat and offer a padded arm to the dog (best use your right arm). Once the dog has taken a grip, use your left hand to grab it by the neck or beat it on the head with a rock or stick. Make sure that whatever you do to the dog, it is permanent; otherwise it will just become even more angered. Under no circumstances should you turn your back on a rabid dog, as it is one way to guarantee being bitten.

Summary

Working as a soldier of fortune can be an exciting life and earn you a lot of money. However, if you are new to soldiering, don't forget to learn as much as you can before you go to your operational theatre. Make sure the company you are working for is signed up to ICoC and that you are not going to become a mercenary in the true scene. The UN is coming down hard on mercenaries, and France has already banned their citizens from becoming one. Stay legal. As regular armies decrease in size, the PSI will expand, which means more contracts offered by governments and more work for the operators. Governments will therefore have to recognize the legitimacy of the PMCs and their right to bear arms just as the regular soldier do, but you can bet your bottom dollar they will seek to enforce tight measures of control. It will be interesting to see what level the PSI has reached in about ten years time. They are already one foot up the ladder, but I wonder if we will ever see a small war contracted out to a PMC.

I have been a soldier or in the military industry since I was seventeen years old; I have fought in wars on just about every continent, and if I have learned anything, it is that the majority of people are good. However, in a war zone, especially one that is governed by a degree of religious fanaticism, you may well be confronted by hatred. My final piece of advice is if you ever feel threatened and it is not within your power to change the state-of-play or improve your odds, then get the hell out.

ANNEX A

This annex is a sample list of just some of the private security and military companies operating around the world. I am not sure that all are still trading, but a quick check of the Internet will verify this. The list provides a good cross selection of different PMCs, providing a whole range of protection and training capabilities. There are literally hundreds more if you search the Internet.

3D Global Solutions. http://3dglobalsolutions.net.p2.hostingprod.com/

AEGIS. www.aegisworld.com/

AGS. www.ags.aecom.com/

AKE Group.www.akegroup.com/

ALGIZ Services, Ltd. www.algiz.eu

Al-Mulla Security Services (Kuwait). www.securicor.com/kw

AMA Associates Ltd [UK]. www.ama-assoc.co.uk/ama2/index.php

Andrews International. http://www.andrewsinternational.com/

Anteon International Corp. www.anteon.com

Armor Holdings. www.armorholdings.com/

ArmorGroup. www.armorgroup.com/

ATCO Frontec Security Services. www.atcofrontecsecurity.com/

Babylon Gates. www.babylongates.com/home.htm

Beau Dietl & Associates. www.beaudietl.com/

Beni Tal (Israel). www.bts-security.com/

BH Defense LLC. www.bhdefense.com/

Blackwater USA. blackwaterusa.com/

Blue Hackle Limited. www.bluehacklena.com/

Britam Defence, Ltd. www.britamdefence.com/

Carnelian International Risks. www.carnelian-international.com

Centigon. www.centigon.com/

Centurion Risk Assessment Services, Ltd. www.centurionsafety.net/

ChaseWaterford. www.chasewaterford.com/

Combat Force (South Africa). www.combatforce.co.za

Combat Support Associates. www.csakuwait.com

Control Risks Group. www.crg.com/

CSS Global. www.gocssglobal.com

Cubic Corporation www.cubic.com/

Diligence Middle East (United States). www.diligence.com/middle-east.html

DynCorp International (United States). www.dyn-intl.com/

Edinburgh International. www.edinburghint.com/?gclid=CIzXquzP6bECFcxofAodWg4A5Q

EOD Technology, Inc. www.eodt.com/

G4S. www.g4s.com/

Garda World Security Corporation. www.gardaglobal.com/

GardaWorld. www.garda-world.com/

Global Options. www.globaloptions.com/

Gpw. www.gpwltd.com/

Granite Intelligence. www.graniteintelligence.com/

Gray Page. www.graypagelimited.com/

Hakluyt & Company. hakluyt.co.uk/

Hill & Associates. www.hill-assoc.com/
Homeland Security Corporation. www.about-hsc.com/
International Consultants on Targeted Security. www.ictseurope.com/
Kroll. www.krollworldwide.com/
Marsh Risk Consulting. www.marshriskconsulting.com/
McNeil Technologies. www.mcneiltech.com/
Ocean5. www.ocean5.co.uk/
Prevent International. www.prevent-international.com/
Securitas AB. www.securitasgroup.com/
SkyLink Security. www.skylinksec.com/
The Brink's Company. www.brinks.com/
The Risk Advisory Group. www.riskadvisory.net/
The Steele Foundation. www.steelefoundation.com/
Vance International. www.vanceglobal.com/
Wackenhut. www.wackenhut.com/

ANNEX B

I have incorporated this annex to provide those with limited experience on how to write a training proposal. I have also added several lesson plans taken from the training manual which I wrote some years ago. They are a little old, but most of the content is still valid today. Together they should provide any PMC member with an idea of what's required should one be required to train other military or police personnel.

COUNTER-TERRORIST TRAINING PROPOSAL

Introduction

At the request of the Military Command of Somewhere, Training International Limited was asked to produce a proposal on the full range of counterterrorist (CT) skills it can offer the Army Special Forces Team to develop and enhance its operational capability. This document covers the training of commanders, assaulters, and snipers, including methods of entry and driving.

Training Phases

It is understood that the army unit will supply forty fit and enthusiastic officers for the training package that will be carried out in Somewhere. The following conditions are recognized but not mandatory:

- Training International will supply three full-time trainers (additional part-time specialists may also be required).
- Duration of the course is six weeks.
- Location is Somewhere.
- All weapons and munitions will be provided locally (list of weapons and equipment to be provided is attached).
- Full course notes to be provided in both English and Arabic for all classroom lectures. Notes in English only for outdoor activities.
- Army Special Forces team will provide all interpreters.
- All CT equipment and personal dress will be provided by Training International. The option to purchase the equipment after training is complete will be granted (i.e., A–Offer Training Price, B–Offer Kit price, C–Offer Complete package training and equipment).
- Army Special Forces team will supply accommodation vehicles, transportation, ranges, weapons, ammunition, and explosive as required.

Each week will be considered a five-day working week. This may change from time to time as facilities and training areas are made available and/or if remedial training is required to enable all officers to achieve the required high standards. It is stressed that this is not a selection process but a dedicated training course with the aim of enabling every officer to achieve his place within the CT team.

Phase 1

During this training phase, all course attendees will focus on the basic skills required for being an effective CT assaulter. This is an essential basis for future training, and it also enables multiple deployments in future training and operations. During this phase and in conjunction with the unit commanders, individuals may be chosen to take on specific tasks as the team progresses towards Phase 2.

PHASE 1 SUBJECTS INCLUDE:

1. Physical training
2. Close-quarter battle (CQB) shooting
3. Options training
4. Sniper and information reporting
5. Support weapon training
6. Methods of entry
7. Medical training
8. Signal training

1. PHYSICAL TRAINING

The fitness training, which is completed daily, is designed to develop agility and speed of movement, and is aimed at developing mental alertness. This training also incorporates unarmed combat to instill the fighting spirit within an elite team, enabling all members to subdue unarmed aggressors and teach them prisoner-handling techniques. The outline syllabus will include:

Strike techniques

Blocks

Baton use

Kuboton and head techniques

Close-in fighting

Punch and draw of weapons

Pressure-point control

Handgun and knife disarm and retention

Ground work

Intervention team skills

The course attendees will develop towards carrying out physical activities wearing full CT equipment, including respirators. These periods of training are progressive to enable all officers to reach a good standard of fitness, climbing ability, and agility.

2. CLOSE-QUARTER BATTLE (CQB) SHOOTING

This phase is the basis for all future training and development. This subject takes up a large part of the overall training as it is crucial that all the officers develop their weapon handling and accuracy to the highest levels to ensure the safety of hostages or civilians within numerous scenarios, whilst accurately firing their weapon systems to achieve the mission objective.

It is emphasized that the correct weapon for the task should be used—a recommended and proven weapons and equipment list is attached. This list has been compiled from experience and universal weapon and equipment trials and is currently in use by the major recognized CT teams worldwide.

Semi-Automatic Pistol:
- Grouping and zeroing
- Application of fire, single and double tap
- Drawing with/without thumb break
- Ball and dummy
- Weak hand firing, distance shooting
- Static drawing turns/other firing positions
- Moving drawing turns/different methods of fire
- Malfunction drills (submachine gun)
- Assessment shoots, competition shoots

Sub-Machine Gun:

Grouping and zeroing
Application of fire, single and double tap

Ball and dummy

Weak hand firing

Instinctive firing

Firing from other positions

Long-range shooting

Static and moving turns

Malfunction drills (with pistol)

Team fire and movement

Assessment shoots, competition shoots

The sub-machine gun (SMG) is the assaulters' primary weapon, more suitable for use within buildings and confined spaces. It is normally fitted with torches and laser sights and is accompanied by three magazines. The pistol is a secondary weapon and should be of the same caliber as the SMG.

3. Options Training

This phase will take up most of the training time, and as it develops, the scenarios will become more complex. Complete rehearsals and currency in live training is essential to maintain all aspects of team safety and efficiency. The subjects covered will develop at the pace of the complete team.

There are four common options that are open to all terrorist incidents:

Option 1. House/Building Assault

Includes methods of entry, room combat, multi-room combat, and full-building combat. Prisoner and hostage handling reception is practiced throughout.

Option 2. Aircraft Assault

Includes methods of entry, aircraft clearing, and prisoner and hostage handling.

Option 3. Vehicle Assault

Car, minibus, bus, coach stopping actions, methods of entry, clearing, and prisoner and hostage handling.

Option 4. Open-Air Option

Plans to react to the terrorist moving hostages or themselves in the open between buildings and vehicles. Alternatives to assault and segregation.

Option 5. Ship Alongside

During this phase the option of "ship alongside" will be added with all variants of scenarios. It must be noted that the option does not cater for a ship or boat at sea: This is purely a land-based operation.

It is during these options that climbing, ladder drills, fast roping, and abseiling will be introduced and practiced.

4. SNIPER AND INFORMATION REPORTING

All members of the team will undergo basic sniper training, which will allow the natural selection of the future CT snipers. It will also ensure that the team members have a sound understanding of the sniper's responsibility and can correctly interpret the information that is reported from the sniper positions.

Sniper Rifle:

> Grouping and zeroing
> Range shooting up to 500 m (547 yd)
> Judging distance

Information Reporting:

> Digital photography and video
> Air-photograph interpretation
> Understanding building color codes
> Accurate observation and reporting techniques

5. SUPPORT-WEAPON TRAINING

All support weapons that are available are incorporated into the CT training. The emphasis in the training is safety, accuracy, and correct utilization. Support weapons required to assist the CT team are:

- Shotgun semi-automatic (Remington Wingmaster folding stock)
- Grenade discharger L1A1 (or similar)
- Arwen 37 mm multi-shot anti-riot or equivalent riot type gun

The listed weapons have non-lethal ammunition available which range from rubber bullets and smoke dischargers to gas-filled

Barry Davies

An assault team using a support weapon to remove door hinges and lock prior to entry.

containers, all of which have many uses within the tactics of a CT team assault.

6. METHODS OF ENTRY (MOE)

This is a crucial part of the CT training as effective entry—along with the element of surprise—can lower the risk to hostages and increase the chance of complete success of any complicated operation. All team members will go through this training during this phase. This not only gives them a complete operational understanding but will also provide valuable assistants to the nominated MOE persons during Phase 2 and in future exercises and operations.

This basic MOE training will cover the following subjects and will be practiced within the Options Training throughout the training phase:

- Forced entry using sledgehammer, crow bars, axes, and hooligan bars
- Removing door hinges and locks with the shotgun
- Safety rules related to handling detonator cord and electric detonators
- Explosive entry using detonator cord on doors and windows
- Explosive safety handling
- Explosive entry using small charges
- Construction and detonation of distraction device

During this phase, the CT team will be introduced to awareness of booby traps and improvised explosive devices.

7. MEDICAL TRAINING

This training is designed to enable team members to carry out immediate first aid on colleagues during training or operations and also to provide the same medical attention to hostages and other casualties.

The outline syllabus of this medical training is:

- Priorities of first aid
- Gunshot wounds and bleeding
- Fractures
- Burns
- Smoke inhalation
- Effects of gas

During the course of the training, an ambulance with trained staff should be available at each training point. If a dedicated ambulance crew is assigned to the CT team, it can then be incorporated into the training exercises.

8. SIGNAL TRAINING

Communication and radio training will be carried out throughout the training. Historically, this highlights the disadvantages of any equipment and ensures these weaknesses, as well as the strengths, are considered and utilized during assault training. The voice procedure adopted emphasizes:

- How to send quick, concise messages
- How to read back accurate instructions
- Coordination of CT operations

SUMMARY OF PHASE 1

Phase 1 is a progressive phase that on completion will have trained an officer to the required level to be a CT assaulter. Throughout the training, the officers will progress from normal uniform handling to wearing full equipment and respirators; this increases operational efficiency in all option scenarios. The officers would be capable of deploying to a local incident at this stage. On completion of this phase, a preliminary team structure will have been identified, in liaison with the commanders, in preparation for moving on to Phase 2 of the training project.

Phase 2

During this phase of training, there will be constant coverage of the skills obtained during Phase 1. This will be in the form of revision, options training, or natural use throughout Phase 2.

The team will start to take the form of an organized unit, and individuals will be allocated positions and responsibilities within the team. The forty men will be split into their command structure: commanders, assault teams, sniper teams, support, and communications.

The training will advance rapidly, utilizing the training areas and public areas. The training will appear complex at times; however, the methods used throughout are based on a tried and tested method of training known as the TEWT system (Training Exercise without Troops). This was designed to "walk and talk" the commanders through an exercise and thus enable them to ultimately provide the strategy or solution to any possible scenario.

PHASE 2 SUBJECTS INCLUDE:

1. Phase 1 subjects
2. Gas-environment training
3. Command and control training
4. Sniper training
5. Method of entry
6. Driver training
7. Helicopter training
8. Options training
9. Key-point visits

1. PHASE 1 SUBJECTS

All the subjects will be revisited throughout Phase 2 regularly, and also assessments will be held on shooting subjects to ensure efficiency, accuracy, and competency is maintained.

2. GAS-ENVIRONMENT TRAINING

Although the officers would have been using their respirators throughout Phase 1, the introduction of working in a real gas environment is essential. It ensures that all respirators are working effectively; it also prevents students from removing the respirators when unsupervised. The training will consist of the following exercises:

Individual respirator check
Individual experience of gas exposure
Operating in a gas environment
Understanding the effects of gas on colleagues and hostages
Fire and smoke survival techniques

3. COMMAND AND CONTROL TRAINING

This subject is primarily for the commanders of the CT team; however, there are lectures that the whole team will attend to understand the workings behind the scenes. It is also worth encouraging supporting agencies or higher commanders to visit the command center during training exercises to understand and develop procedures for future operations.

Call-out and deployment procedure
Liaison with agencies and their responsibilities
CT team holding area
General planning
Emergency response plan
Deliberate attack plan
Hostage negotiations
Post-operations procedure

This particular training area will develop naturally with a question and answer approach: The test subjects will first be gone over as a paper exercise, followed by "walk through, talk through" in the training areas and outside facilities. Then these same scenarios will be practiced with the active CT team.

4. SNIPER TRAINING

Eight snipers will be selected from each of the two teams, making a total of sixteen snipers. Although the sniper training will be their priority, these officers will still be practiced in assault techniques to support the main team players in the event of an emergency response or split plan. The snipers will be chosen by merit of being the best shots from Phase 1 sniper training and by their aptitude for spending long intervals isolated in a confined space. The training will include:

- Camouflage and concealment
- Judging distance
- Stalking, including field craft
- Shooting range 100–1,000 m (109–1,093 yd)
- Firing from cover, including natural, manmade, vehicles, and buildings
- Firing from high-buildings/objects
- Firing from helicopters
- Map reading and aerial-photo interpretation
- Visual intelligence gathering and reporting
- Digital photography and video
- Support weapon

The training will be demanding and variable. The sniper is a very important tool for the commanders, as he or she is normally the main source of information in the identification of terrorist and hostage movements within the incident area.

5. METHOD OF ENTRY (MOE)

The MOE training is carried out by the assault teams; however, the need to identify individuals as the key MOE personnel should be done to give nominated responsibility to maintaining the stores and explosive equipment used. It is suggested that four officers from each team are nominated. The subject is extensive, and techniques are evolved regularly by the user. The equipment available to date, to affect entry either by forced, mechanical, explosive, or covert methods, is extensive. This subject should be given priority as failing to breach a building can have disastrous consequences both politically and operationally. Subjects included in the training are:

Close-target recon

Nonexplosive entry doors and windows

- Sledgehammer
- Hooligan bar
- Bolt croppers
- 1- and 2-man ram
- Hydraulic tools

Wall-breaching cannon

Explosive entry

- Safety rules for explosives and electric detonators
- Construction of internal and external door charges
- Construction of window charges
- Construction of wall charges
- Use of nonlethal wall-breaching equipment
- Use of steel and metal cutters
- Construction and placement of explosive distractions

Covert methods of entry

- Basic lock bypassing
- Alternative methods of entry

6. DRIVING TRAINING

The aim of the driver training is to ensure that the team can be transported by road safely and without incident; ideally all team members should be able to drive. The team needs to be self-sufficient and not

Barry Davies

Sometimes you just have to blow a hole in the wall in order to confuse the terrorists and gain entry.

rely on outside drivers. The main subjects taught and practiced during the training are:

- Convoy driving
- Armored car familiarization
- Fast and progressive driving
- Driving fitted for emergency response
- Using the vehicle as cover during incidents

It is also recommended that a number of officers are trained to drive buses in the event of it being required either to assist with an option or if demands are made by the terrorist to transport hostages to another destination.

7. HELICOPTER TRAINING

The highly trained and respected CT teams around the world have dedicated light helicopters attached to their units. This has the advantage of maintaining common objectives and effecting constant training with the same crews and aircraft. The helicopter is a crucial part of any CT team and gives a wide range of flexibility and options which will become apparent during the training. The helicopter training will include:

- Air movements
- Airborne command post

Hover jumping

> Land
> Buildings
> Ships in dock (ship alongside)

Fast rope techniques

> Land
> Buildings
> Ships in dock (ship alongside)

Abseiling

> Land
> Buildings
> Ships in dock (ship alongside)

Team and casualty extraction

> Buildings

Multi-helicopter team moves

> Remote-site training
> Quick reaction force training

Search and rescue

8. Options Training

The variants of options training, listed in Phase 1, will be progressed and practiced by the teams commanded by their respective leaders.

The options training exercises will be allocated longer periods of time to train the commanders and all individuals in their respective skills. The aim of the final phase of options training is to build up to a complete stand-alone unit with the instructors acting as observers.

Option 1. House/Building Assault

Option 2. Aircraft Assault

Option 3. Vehicle Assault

Option 4. Open-Air Option

Option 5. Ship Alongside

During this phase the option of "ship alongside" will be added with all variants of scenarios. It must be noted that this option does not cater for a ship or boat at sea; this is purely a land-based operation.

During this phase of training, the scenarios will be as realistic as possible. The training should take place outside of the standard training areas, and this could be in the form of different barracks, government facilities, prisons, or disused warehouses.

The training that is done outside of the training area will use blank ammunition and simulated explosives.

It is important to have people acting as hostages to highlight the problems of hostage handling and terrorist identification.

9. KEY-POINT VISITS

Familiarization visits should be arranged during this phase of training to installations within the country which may be attractive targets for terrorists in the future. This is a useful exercise in getting to know these potentially vulnerable areas, but it also serves to establish a planning file, including liaison contacts at each location. The key points should include:

- All major airports
- Seaports
- Conference halls and hotels
- Government buildings
- VIP accommodation
- Ex-pat compounds
- Oil refineries/rigs

A "walk through, talk through" exercise should be completed at as many locations as possible and in the silent hours. Contingency plans can be drawn up in preparation for a real incident to give the team a good starting point in planning holding areas, advantageous sniper points, a potential command and control center, best routes in and out, and preplanning the position for such things as an aircraft under duress and where it should be parked to the CT team's advantage, etc.

SUMMARY

As you can see, each part of the training has been broken down into the different skills required to be taught. The next thing is to make sure that the instructors who will be doing the teaching are all singing from the same hymn sheet. The best way to do this is to have a standard set of "lesson plans" covering each subject. You will need to write out a detailed lesson plan for both classroom lessons and practical training. The reason for doing this is fairly simple: It offers consistency in training. If one instructor is ill or not available, another can stand in, pick up the lesson plan, and teach the students. A sample lesson plan is outlined below. Typically, this would be made into a PowerPoint presentation with added images to qualify what the instructor is saying.

Training Division Lesson Plan

COMMAND & CONTROL
Media Interview

Author: Barry Davies   **Course Director**

	BEGINNING OF LESSON	
PRELIMS	Introduce yourself and the lesson. During this lesson there will be no smoking, eating, or drinking. In case of a fire, leave the building via the fire exits and muster at the designated area. Do not touch anything on your desks until told to do so. There will be a requirement to take notes. Question policy: Pose, pause, nominate. The reference to this lesson is part of the BCB Counter-Terrorist Manual	
REVISION	Nil.	
INTRODUCTION	Counter-terrorist measures, such as an assault on an occupied building or an anti-hijack, almost always demand the public's attention. Although interviews with counter-terrorist team members are rare, there are times when senior commanders are required to address the public to answer questions about the operation.	
INCENTIVE	Use it to your advantage to give a formal interview stating clearly the facts as they occurred. This can be particularly useful when deaths have resulted from an assault.	
OBJECTIVE	The objective is to teach you some simple "do"s and "don't"s when confronted with a media interview.	

Barry Davies   **Course Director**

	MIDDLE OF LESSON	
	Successful Interview Techniques Understandably, most people are apprehensive about saying the wrong thing . . . especially if there have been a number of civilian casualties during any assault. The thought of being interviewed can unnerve even the most stouthearted. They've heard about or even experienced the outcome of provocative questioning by a reporter or talk show host. They believe that by being aggressive they will get answers that are interesting to their audiences. With adequate preparation and some understanding of the news media, you will not be at a disadvantage when handling this type and other media interviews. This guide provides practical advice on preparing for and protecting yourself during interviews, on using interviews to your advantage, and on how those inevitable preinterview jitters can be put to work for you. It is almost impossible to overprepare for a media interview.	
	In Advance • Set Goals: Before doing an interview, define your purpose for doing it. What do you intend to accomplish by doing the interview? Write these down. • Define Key Points: Prepare a short list of key points to be made in the interview. Limit these to two or three for most interviews. • When the media calls you, before confirming the interview, ask for the name and phone number of the person who will be doing the interview, the purpose of the interview, who else will be interviewed, and when the story is likely to appear. Keep a record, especially if multiple interviews are involved. • Preparation Time: Allow for adequate preparation time in advance of an interview. Spontaneous interviews can be dangerous. Going into an interview unprepared can make you appear indecisive or evasive, even incompetent.	

- Q&As Prepare a series of possible questions that could be asked, and then prepare concise answers. Give priority to the most awkward and potentially difficult questions a reporter could possibly ask.
- Never Refuse: Refusing to do an interview is almost never a reasonable option. You can be sure the information vacuum created by a refusal will be filled, likely with comments from those less well-informed.
- Be Comfortable: Be sure the interview room is made comfortable for you. That means a comfortable, upright chair, room temperature at or slightly below normal (TV lights will heat the room), and a side table for a glass of water and any necessary papers. Avoid placing a table or desk between you and the reporter; it creates an impression of aloofness, and avoid swivel chairs for TV interviews. Most people want to swing back and forth—it makes you look nervous. For a press conference, different arrangements will be needed.
- Avoid Rushed Interviews: Always avoid being stopped at the incident site and asked for an interview—wait until you are prepared.

Avoid giving impromptu interviews.

The Interview

- Never Off The Record: There is no such thing as "off the record." Assume everything said within the hearing of a reporter will be used. Unguarded comments can be embarrassing. A favorite trick of some reporters are "warm-up questions" before the interview, or appearing to turn off a camera or tape recorder afterwards, then asking "informal" questions, implying these are off the record. They aren't.

- Be Brief: Keep answers crisp and brief. One or two sentences usually suffice. Let the reporter keep asking questions. It's the interviewer's job. Voluminous and complex answers won't make it to the news report and may lead to a line of questioning away from your objectives.

- Never use "No Comment": It's considered rude. Worse yet, it will be taken as a signal you're hiding something. A better answer would be: "We don't have enough information on that right now to give you an accurate answer. Once we have a better understanding, I will be pleased to share it with you."

- We and Us: Use the term "we" rather than "I" wherever possible. The word "we" suggests you belong to an organization. Whenever possible, link your answers to the public interest rather than just yours or the organization's.

- Ignore the Audience: If an audience is present during the interview, ignore it. Focus on the person asking the questions. Similarly, avoid looking at the audience.

- Eye Contact: Maintain eye contact as much as possible with the interviewer. Eye contact helps build a rapport and will ensure your picking up on cues about the interviewer's temperament and line of questioning.

- Stay Cool: Always remain polite and calm. Questioning can become intensive, sometimes aggressive. This is common. But an outburst of anger or frustration will almost certainly become the story, diverting attention from the real subject. Try to keep the interview a conversation, not a confrontation.

- Expect Courtesy: People being interviewed have every right to be treated with courtesy and respect. If you feel you are not being treated properly, say so. If the objectionable behavior persists, terminate the interview politely. Remember, questions that make you uncomfortable may not necessarily justify terminating the interview.

- Avoid Offensive Remarks: If a question contains offensive language or remarks, avoid at all costs repeating the question, e.g., "you and your team are just a bunch of killers." A denial may be the only direct quote used; the result could imply the opposite of what you intended. Keep your answer positive. Better to say something like, "Contrary to what you may have heard, the counter-terrorist team is trained at the highest professional level, and we as a country should be proud of it."
- Correct Errors: Similarly, you should correct any false information that is contained in a question, but never repeat the false information. Your answer could appear in the news report, giving credence to the false information. Simply begin your answer with: "Apparently you have been given inaccurate information. The correct information is . . ."
- Avoid Jargon: Keep the language in laymen's terms. Otherwise, people outside the industry will have no idea what you're saying.

Be relaxed and project authority.

- Be Truthful: Even if it's embarrassing. Answer embarrassing questions directly and briefly, and then move on. Remember, untruths will almost certainly be unmasked. Your credibility and that of organization will suffer. The public may not like bad news; they will accept it, but they most certainly will not tolerate dishonesty.

- Complex Questions: Interviewers sometimes ask multiple-part questions. These are opportunities. Begin by saying: "Your question contains a number of questions. I'll answer them one at a time." Answer first the portion you are most comfortable with. This gives you time to think about the others, particularly the more difficult ones. Often the reporter will move on before you get through all of the questions.
- Be Yourself: Some people try to adopt a more "authoritative" or "professional" appearance. Don't. Being yourself is the best guarantee you have of being at ease in the interview. You will come across as genuine.
- Biographies: Provide a brief biography of the counter-terrorist team. Emphasize the members' qualifications and experience relevant to the subject of the interview.
- Confidence Wins: Let your self-confidence show. Your confidence will inspire others to have confidence in you and your team. But overconfidence can be interpreted as arrogance and will create a barrier between you and your audience.

- Shake hands with the Interviewer: This shows that you hold no malice even if the interviewer was harsh with you.

Avoid using hand gestures.

After the Interview

The period immediately following an interview is a risky time for the unwary. The tendency is to relax and let your guard down. Reporters know this, and some will ask questions that appear to be harmless. They are not. Here are a few points to keep in mind:

- Watch Out!: The TV lights may go off, but the camera and sound recording may not.
- Please Don't Ask: It is considered impolite to ask reporters to see their stories before they are printed or aired. They will regard your request as an insult. Most will find it difficult not to carry that negative feeling while preparing their news story about you and your organization.
- Getting Corrections: If the news report is substantially inaccurate, ask for a correction. Review the news reports thoroughly before doing anything. Make certain the inaccuracy is of consequence. The audiotape of the interview will come in handy. The first call should always be to the reporter. Avoid a confrontation. It is possible the reporter will have misunderstood something and will want to put it right. If the reporter is unwilling, talk with the reporter's editor. Chances are, the matter will be resolved. The error will have to be quite substantial, in their opinion, for them to issue a correction.

	ANY QUESTIONS ON THE ENTIRE LESSON	
SUMMARY	Counter-terrorist actions almost always have a certain amount of "daring," of which the general public has an interest. However, operations can also produce casualties, such as a dead terrorist or worse, dead hostages. A media interview provides the CT team with a voice.	
LOOK FORWARD	Your next lesson will be an in-studio practical session, which will be recorded for analysis.	

Training Division Lesson Plan

COMMAND & CONTROL
Lock-Picking

Author: Barry Davies   **Course Director**

		BEGINNING OF LESSON	
	PRELIMS	Introduce yourself and the lesson. During this lesson there will be no smoking, eating, or drinking. In case of a fire, leave the building via the fire exits and muster at the designated area. Do not touch anything on your desks until told to do so. There will be a requirement to take notes. Question policy: Pose, pause, nominate. The reference to this lesson is part of the BCB Counter-Terrorist Manual	
	REVISION	Nil.	
	INTRODUCTION	It is often the case during any CTR or assault that a covert method of entry is required. Lock-picking enables us to enter through a locked door or window as if we had a key.	
	INCENTIVE	Lock-picking is a special skill which is properly attained only by years of constant practice.	
	OBJECTIVE	My objective is to teach you basic lock-picking skills so by the end of the lesson you will be able to understand how a lock works and how it can be opened using a variety of tools other than the key.	

Barry Davies   **Course Director**

	MIDDLE OF LESSON	
	Lock-Picking All counterterrorist units must learn the basics of lock picking as part of their MOE skills. The principle of picking locks is fairly basic, as are the lock-picking tools, many of which can be easily made or improvised. The problem with picking locks lies in the skill. It can take many years to perfect the fundamentals of lock-picking, and constant practice is required in order to acquire the "feel." Outlined here is a very brief guide into the lock-picking process.	
	Disclaimer Unless carried out as a professional trade, carrying lock-picking tools in public is illegal. Mention of lock-picking is made here solely in the context of counter-terrorist techniques.	
	PIN TUMBLER LOCK Most locks that have been manufactured over the past twenty years are of the pin tumbler type. In its basic form it is a very simple locking device. A series of small pins fit into the inner barrel of a cylinder. The pins are split in the middle, normally at different lengths, and are forced into recesses within the inner barrel by a small spring. If a correct key is inserted, the different-sized pins are brought into line were their splits meet the outer casing of the inner barrel. This allows the inner barrel to turn freely within the casing and thus release the lock.	

A Typical Lock-Picking Kit Issued to Counterterrorist Units

Any method of aligning the pins in this manner, and turning the inner barrel with open the lock, this can be achieved by two methods: racking or picking the pins. This requires two basic tools, a lock pick/rake and a tension bar.

The pick or rake is a simple flat strip of hardened metal that has its end shaped to fit into the lock and advance the pins on their small springs to the required depth.

The tension bar is a simple flat strip of metal, inserted into the mouth of the barrel in order to employ a minute amount of tension onto the barrel; this process helps to seat the pins and to turn the barrel.

Note: There are many different types and designs of lock-picking tools that can all be used for different functions. I suggest that two are sufficient.

RAKING

Raking is the quickest method of opening a lock. It is fast and easily done, providing the pin sizes do not change suddenly, such as the combination illustrated below. The lock should be clean and free from any grit or dirt; blowing hard into the lock before attempting to open it is a good idea. Raking is simply a matter of inserting the pick to the rear of the pins and swiftly snapping the pick outwards, running the tip over the pins in the process.

Rake from the rear by rapidly removing the ripple pick.

Prior to doing this, a tension bar is inserted into the bottom of the keyway, and a slight pressure is applied on the inner barrel of the lock. The tension is applied in the unlock direction. The amount of tension should be just enough to turn the barrel once the pins are seated, but not so strong as to bind the pins against the barrel. It is this single "feel" that is the basis of all good lock-picking. If the tension is too heavy, the top pins will bind and the sear line will not allow the breaking point to meet. If the tension is too weak, the pins will simply fall back into the locked position.

When raking, it will be necessary to repeat the operation several times. If the barrel does not turn by the fourth time, hold the tension in place with the tool. Place your ear to the lock and slowly release the tension; if you hear the pitting sound as the pins fall back to rest, then you have applied to much pressure, if you hear nothing, then you need to apply more pressure on the tension bar.

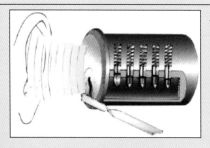

Release the Tension Bar and Count the Pins

The ease at which a lock can be opened will depend on three things: firstly, the length and position of the pins; secondly, the type of tools you use; and thirdly, the manufacture of the lock. Cheap locks will be easier to open than expensive ones. Cheaper locks are generally poorly constructed, allowing a much greater clearance between the barrel and the body, which allows for easy assembly during manufacture. Another point of cheap locks is poor barrel alignment and oversized pin holes; both these points make it very easy to pick the lock.

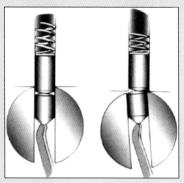

	When Raking Fails, the Pins Need to Be Picked Individually	
	Lock-picking is very similar to raking but requires a lot more skill as the pins need to be seated individually. Staring at the back of the lock, feel for the rearmost pin and gently push it up; the barrel should move a minute fraction. Working towards the end of the lock, seat each pin in turn until the barrel is released. A combination of one swift rake followed by picking is sometimes the easy answer.	
	Note: It is also worth sharpening one end of your pick to a needle point. If this point is forced all the way to the rear of the lock until it hits the rear plate, the sharpened pick will grip the metal. Try forcing the plate down or up; this will sometimes release the lock without the need for raking or picking.	
	Show Lock-Picking Movie (BCB CT Training Lessons – 2)	
	ANY QUESTIONS ON THE ENTIRE LESSON	
	END OF LESSON	
SUMMARY	To summarize, it is essential that you can identify the correct type of lock and that you use the correct tools to open it. Touch and feel are the most important aspects of lock-picking.	
LOOK FORWARD	In the next lesson you will be using improvised tools to open various locks as well as looking at electric lock-picking equipment.	

Training Division Lesson Plan

COMMAND & CONTROL
Room Combat

Author: Barry Davies   **Course Director**

	BEGINNING OF LESSON	
PRELIMS	Introduce yourself and the lesson. During this lesson there will be no smoking, eating, or drinking. In case of a fire, leave the building via the fire exits and muster at the designated area. Do not touch anything on your desks until told to do so. There will be a requirement to take notes. Question policy: Pose, pause, nominate. The reference to this lesson is part of the BCB Counter-Terrorist Manual	
REVISION	Holding Area Approach MOE	
INTRODUCTION	One of the most important aspects of counterterrorism is room combat. While the mechanics of room combat are fairly simple, it is vital that the basic drills are adhered to in order to clear the room and minimize casualties.	
INCENTIVE	Entering a room which has an armed enemy presence is one of the most dangerous aspects of counter-terrorist work. Good room combat skills minimize the risk to both team members and hostages.	
OBJECTIVE	Therefore, it is my objective to demonstrate to you the basic room combat skills based on a single-room technique.	

Barry Davies		Course Director
	MIDDLE OF LESSON	

Single-Room Combat Using a Four-Man Assault Team

In this lesson we will examine the procedures for a four-man assault team entering a room that has a single door. While it is recognized that some doors will require a breaching team (see MOE), for this example we will ignore this aspect. Based on the assault plan, the team leader designates the assault team and identifies the location of the entry point for them. In some cases, especially when there is an outside threat, the team leader may also allocate a second team to act as covering support.

Assault team members move with stealth, positioning themselves as close to the entry point as possible. If an explosive breach is required, then the assault team remains in a covered position until the breaching team is ready. When approaching the point of entry, the team must avoid any verbal signals or radio traffic, as they may alert the enemy.

The door is the main entry point into a room. It is also the most vulnerable point of entry as any enemy in the room will concentrate his fire on it. It is essential that team members clear this funnel as quickly as possible.

The team should position themselves outside the door in such an order that they can open the door, deliver distraction ordnance, such as stun grenades, and make a clear entry while avoiding enemy fire. Talking should be avoided, and communications should take the form of a physical squeeze rather that a tap, as accidental dumping could be misconstrued. When the "Go" is given, the team must enter the room in such a way that they accomplish the following:

- Completely dominate the room.
- Eliminate any threat.
- Mark the room as cleared.

If the room size is unknown and upon entry it is found to be small, then the number 1 entry man can call out "short room," which allows for the room to be cleared by the number 1 and 2 entry men only.

In reality, rooms and entrance doors seldom fall into the ideal structure; however, for this exercise we assume that the room is 4 m (13 ft) wide by 3 m (10 ft) deep with the door in a central position, opening inwards and to the left. These simple facts alone dictate the initial movement of individual team members when entering the room. For example:

Each team member has a designated sector of fire assigned to him or her. This avoids any team member crossing the path of another.

The allocated sector of fire is achieved by taking up a dominating position.

A dominating position is determined by the size of the room, objects such as furniture, enemy fire, and hostages.

As in all combat situations, the clearing team members must move tactically and safely. This means holding the primary weapon in the "ready" position, moving with the muzzle pointed in the direction of travel. The weapon butt should be in the shoulder, with the muzzle slightly down to allow unobstructed vision. Both eyes should be open, with the weapon muzzle turning in conjunction with the head. The trigger finger should be on the trigger, with the thumb on the safety sector, which is set at "Safe." The weapon should only be switched to fire just prior to the "Go." Room combat is best carried out with short bursts of three to four rounds.

In order to perfect their room combat techniques, team members must enter the room in a given order, penetrate to their point of domination, and eliminate any threat in their sector. Should any of the team members encounter an enemy threat prior to reaching their dominating position, they should deal with this by shooting on the move, as the room can only be made secure once all points of domination have been reached.

The number 1 man should always be the most experienced team member next to the team leader. The number 1 man is responsible for frontal and entry/breach point security.

The number 2 man is directly behind the number one man in the order of movement

The number 3 man is normally the team leader and is responsible for initiating all voice and physical commands.

The number 4 man normally enters the room last and is responsible for entry point security once entry has been made. Where multi-room clearance is required, he is also the "link man" to other room clearance teams.

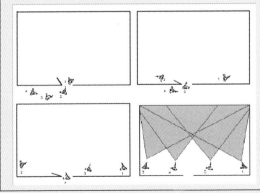

From the moment the door is open (in this case, inwards and to the left) the number 3 throws a stun grenade. This is immediately followed by the number 1 and 2 men making an entry. As the door opens to the left, number 1 will go to the right, with number 2 going to the left. Both 1 and 2 move at the same time, with their main focus of concentration on the area directly to their front, then along the wall on either side of the door.

The number 3 and number 4 men start at the center of the wall opposite their point of entry and clear to the left if moving toward the left, or to the right if moving toward the right. They stop short of their respective team member (either the number 1 man or the number 2 man).

All team members move toward their points of domination, engaging all targets in their sectors. Team members must exercise fire control and discriminate between enemy and hostages. Shooting is done without stopping, using reflexive shooting techniques.

Stoppage: If an assaulting team member has a weapon malfunction, he will immediately adopt the stoppage drill by going down on one knee. This will indicate to other team members that he has a problem, and they can cover his arch of fire. Before rising to his feet, the team member must have cleared the stoppage or drawn his secondary weapon. This drill must be continually practiced until it is second nature.

Unless restricted or impeded, team members stop movement only after they have cleared the door and reached their designated point of domination. In addition to dominating the room, all team members are responsible for identifying possible escape holes, such as an internal door and windows.

Upon domination of the room, the team leader will identify any other immediate threats and direct the appropriate team member to deal with it. Once the situation is secure, the team leader will indicate team members to carry out the following actions:

	<ul><li>Remove weapons from within enemy reach.</li><li>Bind hands of both live and dead enemy.</li><li>Bind hands of all hostages and remove via exit plan.</li></ul>Once the team leader is satisfied that the room is clear, he will call, "Room clear," and inform the team commander. The team then extract themselves via the door in reverse order, finally marking the room as "clear" with a chemical light-stick or any other recognized marking method. **Show Room Combat Movie (BCB CT Training Lessons – 2)** Marking methods: Red – Indicates the main entry point. Green – Indicates that a room is clear. Yellow – Indicates that a medic is required. Blue – Indicates that a room is booby trapped.	
	ANY QUESTIONS ON THE ENTIRE LESSON	
	END OF LESSON	
SUMMARY	It is essential that you learn and practice the basic room combat drills.	
LOOK FORWARD	Your next lesson will be a demonstration of basic room combat; this will be followed by several practical lessons in multiple-room combat.	

ANNEX C

Deaths of Civilian Contractors in Iraq and Afghanistan

I have added this small section to help you make up your mind about becoming a mercenary. The information below takes into account all forms of contract work in both Iraq and Afghanistan. Note that the amount of contractors killed in vehicles, both in IEDs and in accidents, is a very large proportion. Spend a few moments reading through this, as it is a good indication of what could happen to you should you decide to become a soldier of fortune.

Outsourcing work in high-risk areas and war zones, especially those jobs previously done by soldiers, is a prelude to certain death for many. As of January 2012, there were 100,000 defense contractors (of which some 25,000 were American) operating in Afghanistan, compared with 90,000 American soldiers in the same region.

In 2011, 430 employees working for American contractors were killed in Afghanistan, of which 386 were working for the Department of Defense. By comparison, 418 American soldiers died in the line of duty in Afghanistan during 2011. It would seem most of the contractor deaths came from those companies providing drivers, security guards, and interpreters.

The trend is similar to that seen in Iraq, where contractor deaths also outnumber those of the military. One of the worst companies to suffer both in Iraq and Afghanistan is the giant defence communications company L-3. Over the past ten years, L-3 and its subsidiaries have lost at least 370 workers, and some 1,789 have been seriously wounded.

Death can come from any direction. In July 2012, three contract workers were killed when a man wearing an Afghan military uniform opened fire at a police training facility in Herat, western Afghanistan. The men were part of the International Security Assistance Force (ISAF) and were simply described as "trainers."

As I stated earlier in the book, you are not safe in a war zone until you are actually on the aircraft and have left the country. In April 2011, nine Americans we killed by an Afghan Air Force pilot at Kabul Airport. It is unsure how the incident started; some say the pilot was suffering from mental illness, had been recruited by the Taliban, or simply had been in an argument with the Americans. Whatever the reason, they are dead. The incident took place in the operations room of the Afghan Air Corps, and when the firing began, there was no stopping it until the pilot was killed. The Taliban claimed responsibility. It's not just the PMC operators that get caught out. On April 7, 2004, two Germans, Tobias Retterath and Thomas Hafenecker, were killed by Iraqi terrorists in an ambush near Fallujah. They were members of the elite counter-terrorism unit GSG-9, working at the German embassy as guards. The second officer, Thomas Hafenecker, is still missing today.

The dangers of becoming a mercenary or soldier of fortune are not restricted to simply stepping on an IED or being shot. Many contractors have been kidnapped and held for ransom. One such contract was beheaded. Just imagine what that person was thinking as the perpetrators of this deed ripped the knife across his throat and struggled for several minutes to remove his head. This is the harsh reality of working for money and being a soldier of fortune.

However, many deaths occur through no fault of the enemy. Major James Stenner of the Welsh Guards and Sergeant Norman Patterson from the Cheshire Regiment died in an accident in Baghdad on January 1, 2004—both were members of a British SAS unit operating in the area. It is reported that they had been celebrating the festive season and were driving close to the "Green Zone" when their vehicle is believed to have struck a concrete barrier forming part of a security chicane.

INDEX